24 Advances in Biochemical Engineering

Managing Editor: A. Fiechter

Reaction Engineering

With Contributions by
H. Binder, K. Buchholz, W. D. Deckwer,
H. Hustedt, K. H. Kroner, M.-R. Kula,
G. Quicker, A. Schumpe, U. Wiesmann

With 99 Figures and 44 Tables

Springer-Verlag
Berlin Heidelberg GmbH 1982

ISBN 978-3-662-15336-9 ISBN 978-3-540-39467-9 (eBook)
DOI 10.1007/978-3-540-39467-9

Originally published by Springer-Verlag Berlin Heidelberg New York in 1982
Softcover reprint of the hardcover 1st edition 1982

Library of Congress Catalog Card Number 72-152360

2152/3020-543210

Table of Contents

Gas Solubilities in Microbial Culture Media

Adrian Schumpe and Gerd Quicker
Institut für Technische Chemie, Universität Hannover, D-3000 Hannover 1, FRG

Wolf-Dieter Deckwer
Fachbereich Chemie, Universität Oldenburg, D-2900 Oldenburg, FRG

Available information on gas solubility in microbial culture media is reviewed. Emphasis is given to oxygen and carbon dioxide solubilities. Experimental techniques which can be successfully applied to culture media are presented. All the parameters which affect gas solubilities, i.e., above all the composition of the media are thoroughly discussed. In general, gas solubilities in nutrition and cultivation media can be predicted by a log-additivity approach. To this end knowledge of the composition of the media and the solubility parameters (K_I) of the individual compounds is required. For a variety of substances encountered in cultivation broths the parameters K_I for oxygen could be evaluated from literature data and are summarized in this paper. Appropriate recommendations for applying direct and indirect predictive methods are given. Cases of failure are mentioned as well.

1 Introduction

The solubilities of gases in liquids are fundamental physicochemical data. They are often referred to as physical or saturation solubilities. In biosciences as, for instance, in fermentation technology, algae cultivation, marine technology, waste water treatment, physiology, and environmental sciences, it is the knowledge of the solubilities of oxygen and, to a lesser extent, of carbon dioxide which is particularly needed. In general, gas solubilities are required

(i) to establish mass balances,
(ii) to calculate yield coefficients (stoichiometry),
(iii) to determine volumetric mass transfer coefficients,
(iv) to design and scale up bioreactors.

Solubilities are essentially responsible for the value of the driving concentration difference of mass transfer between the gas and the liquid phase. It can be assumed that exact knowledge of O_2 and CO_2 solubilities in biological culture media may contribute to obtaining a deeper insight into and better interpretation of various bioprocesses.

In biotechnology, it is especially the O_2 solubility which is of major importance. O_2 determinations in microbial culture media are conveniently carried out with the help of the polarographic probe [1,2]. Such probes give a current which is proportional to the diffusional flux of oxygen across the probe membrane. The diffusional flux, in turn, depends on the chemical potential of oxygen and hence on its fugacity which, for the sake of simplicity, is often called oxygen partial pressure. It is important to note that in aqueous solutions of various composition the oxygen fugacities are equal if these solutions are in equilibrium with the same gaseous phase. However, the actual amount of O_2 present in the solutions may be quite different, of course. This is shown schematically in Fig. 1. The addition of electrolytes and alcohols changes the O_2 solubility while the fugacity and partial pressure, respectively, remain constant. The partial pressure, p, and the solubility, c, are interrelated by Henry's law:

$$p = H_m c_m \qquad (1)$$

where H_m is Henry's constant.

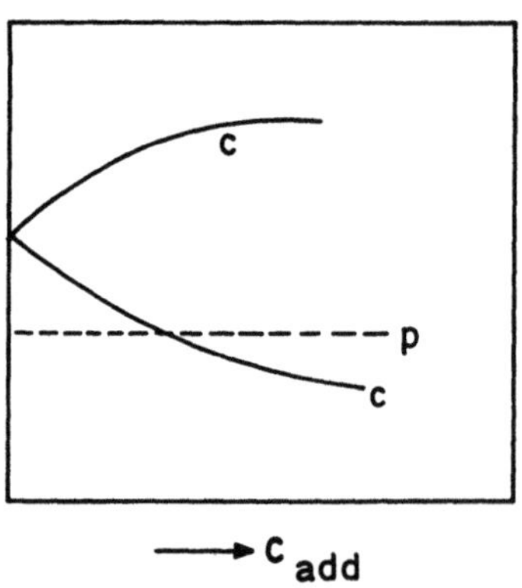

Fig. 1. Principal dependency of partial pressure and dissolved gas concentration on the concentration of added compounds

Numerous experimental data on gas solubilities, particularly for O_2, are available from the literature. However, most data refer to simple systems like solutions of single salts or organic compounds. In contrast, microbial culture media are complex since a number of organic compounds, salt mixtures and metabolites are usually present. The addition of salts and sugars usually decreases the gas solubility while the latter may increase if short-chain alcohols are present. The effect of metabolites on the O_2 solubility seems to be complex but only insufficient data is available.

This paper summarizes knowledge published on gas solubilities in microbial culture media. Emphasis is placed on the solubilities of oxygen and carbon dioxide. Experimental techniques for measuring gas solubilities are outlined and the shortcomings of some theoretical approaches are mentioned. A review is given on the influence of various substances on gas solubility. It can be assumed that many substances dissolved in water and possibly present in biological media act independently. It will be shown that under this condition their effects on gas solubility can often be accounted for by an additive approach. This makes it possible to reliably estimate O_2 solubilities in nutrition media and fermentation broths provided their composition and the individual solubility parameters of the substances present are known with sufficient accuracy. Such solubility parameters evaluated from previous measurements and from literature data are given in this paper and their application is demonstrated. Also other direct and indirect methods for solubility estimations in microbial culture media are discussed.

2 Methods for Expressing Gas Solubility

There are various ways of expressing gas solubility [3]. The equilibrium liquid phase concentration is often given, e.g., as

c mass concentration (mg l^{-1})
c_m molarity (mole l^{-1})
w mass fraction (—)
x mole fraction (—)

stating both the temperature and the pressure the data refer to. If Henry's law holds it may be more convenient to calculate Henry's constants using any of the various concentration measures:

H Henry's constant, e.g.,

$$H_m = \frac{p}{c_m} \ (\text{kPa l mole}^{-1}) \tag{2}$$

$$H_x = \frac{p}{x} \ (\text{kPa}) \tag{3}$$

$$H_L = \frac{c_g}{c} \ (—)\,. \tag{4}$$

The reciprocal of H_L, i.e. the ratio of the liquid to the gas phase concentration, for low solubilities is equivalent to the Ostwald coefficient:

L Ostwald coefficient (—)
(= ratio of the volume of gas absorbed per unit volume of the solvent).

More often the solubility data are corrected for a standard pressure the correction usually assuming ideal gas behavior and validity of Henry's law resulting in one of the following coefficients:

α Bunsen coefficient (—)
(= volume of gas reduced to 0 °C and 101.3 kPa (1 atm) absorbed by unit volume of solvent at a gas partial pressure of 101.3 kPa),

β absorption coefficient (—)
(= volume of gas, reduced to 0 °C and 101.3 kPa (1 atm) absorbed by unit volume of solvent at a total pressure of 101.3 kPa including the solvent vapor pressure).

S Kuenen coefficient ($cm^3\ g^{-1}$)
(= volume of gas (in cm^3) reduced to 0 °C and 101.3 kPa (1 atm) dissolved in the quantity of solution containing 1 g of solvent, i.e., S is proportional to the gas molality)

C_w weight solubility (mole g^{-1})
(= moles of gas dissolved per gram of solvent at a gas partial pressure of 101.3 kPa (1 atm)).

For convenience, the conversions of the different coefficients into the Bunsen coefficient, α, which will be used in this paper are listed below. (For symbols and units see nomenclature)

$$\alpha = \frac{273.15}{T} L \tag{5}$$

$$\alpha = \frac{101.3}{101.3 - P_s} \beta \tag{6}$$

$$\alpha = \varrho_s(1 - w)\, S \tag{7}$$

$$\alpha = \varrho_{st} V_0 C_w \,. \tag{8}$$

In case of low solubilities the following relationships are as well approximately valid:

$$\alpha = \frac{101.3 V_0}{10^6\, p M_G} c = \frac{101.3 V_0}{10^3\, p} c_m \tag{9}$$

$$\alpha = \frac{\varrho_s V_0}{M_G} w = \frac{\varrho_s V_0}{M_{st}} x \tag{10}$$

$$\alpha = \frac{101.3 V_0}{10^3} H_m^{-1} = \frac{p \varrho_s V_0}{M_{st}} H_x^{-1} \tag{11}$$

$$\alpha = \frac{273.15}{T} H_L^{-1} \tag{12}$$

3 Experimental Techniques

3.1 General Methods

Gas solubilities have been measured quantitatively for almost two centuries now and many techniques and apparatuses have been developed for this purpose. There are two groups of methods: chemical methods to analyse gas saturated solutions and physical methods which mainly determine either the amount of gas necessary for saturating the initially gas-free solvent (saturation methods) or the gas which can be desorbed from the saturated solution (desorption methods).

Chemical determinations are essentially gas specific. The most important one of the chemical techniques is Winkler's method for dissolved oxygen analysis. The basic chemical steps involved are the oxidation of manganous hydroxide by the dissolved oxygen in an alkaline solution, reduction of the produced manganic hydroxide by iodide upon acidification and titration of the liberated iodine with a thiosulfate solution. Some of the most precise determinations of the oxygen solubility in water have been carried out by refinements of the fundamental procedure. However, other solutes, e.g., buffering, oxidizing or reducing substances may interfere with certain steps. In some cases, this can be overcome by special modifications [4] but generally only at the cost of both convenience and accuracy.

There are also some physical methods to investigate the dissolved gases directly (e.g., photometric methods) but more often the gas is removed from the solution to avoid possible interferences with other solutes. The desorption may be accomplished by applying a vacuum as, e.g., with the classical van Slyke apparatus. Preferably the gas is stripped from the solution by an inert gas purified and either reabsorbed in pure water for analysis or analysed in the gas phase, e.g., by means of a gas chromatograph or a mass spectrometer. In this way the amount of different dissolved gases can be investigated simultaneously.

The saturation methods also require a preceding desorption process in order to free the liquid from dissolved gases. This is usually accomplished by boiling the liquid under vacuum. Other techniques used less frequently are spraying the liquid through a fine nozzle into an evacuated chamber or evacuating the frozen solution. The latter method reduced the solvent losses involved. Solvent evaporation may be critical especially in the case of mixed solvents since different volatilities of the components result in a change in composition. The gas-free solvent is then brought into contact with the gas. The amount of gas absorbed in the equilibration process may be investigated by observing either the isobar volume reduction or the isochor pressure decrease.

Detailed descriptions and the particular references can be found in the reviews of Markham and Kobe [5], Battino and Clever [3], Clever and Battino [6] and the textbook of Hitchman [1]. Here only a few methods which have so far been applied to microbial culture media will be discussed in more detail.

3.2 Methods Applied to Microbial Culture Media

Gas solubility measurements in microbial culture media have almost exclusively been devoted to the solubility of oxygen. Winkler's method was found to be ill-suited for this purpose due to interferences with other solutes [7,8]. Phillips and Johnson [7] therefore applied a simple calibration technique to obtain a conversion factor for the partial pressure readings of their oxygen probe. A known quantity of oxygen was dissolved in a known quantity of oxygen-free medium while recording the increase in oxygen tension. Since the actual oxygen concentration was known Henry's constant could be calculated. It is not clear, however, how the respiration of the organisms (if present) was inhibited. A similar calibration technique has recently been suggested by Käppeli and Fiechter [9]. They used the decomposition of H_2O_2 catalysed by catalase to generate defined oxygen concentrations. For pure water and solutions of glucose, sodium chloride and a salt mixture the method was demonstrated to result in reasonable solubility values. For solutions containing methanol or ethanol the probe readings turned out not to be proportional to the added amount of H_2O_2 possibly due to some interference of the alcohol with the probe membrane.

A calibration procedure based on the respiration rate of the living microorganisms was applied by Liu et al. [8] in culture media of *Thiobacillus ferrooxidans*. The rate R of the decrease in oxygen tension in the medium without aeration was recorded with an oxygen electrode. Constancy of $\Delta p/\Delta t$ was observed down to a critical oxygen tension of 10 Torr. Therefore, dividing R by the pseudo-steady-state oxygen uptake rate, r, measured in a Warburg-type apparatus Henry's constant H_c was evaluated from

$$\frac{R}{r} = \frac{\dfrac{dp}{dt}}{\dfrac{d(\text{mass } O_2)}{V_L\, dt}} = \frac{p}{c} = H_c \tag{13}$$

A technique for oxygen probe calibration during exponential growth in an agitated bioreactor has recently been suggested by Lehmann et al. [10]. The change in dissolved oxygen tension resulting from a step change in agitation speed at a constant air flow rate is recorded. The corresponding amount of oxygen absorbed into the medium (or desorbed from the medium) is evaluated from the resulting peak in the oxygen exit gas concentration which is continuously measured using a paramagnetic oxygen analyzer. This dynamic procedure obviously requires a low aeration rate in order to obtain sufficiently large effects in the exit gas concentration and, furthermore, the growth pattern of the microorganisms must not be affected by the change in agitator speed. The technique has been applied to culture media of *Myxococcus fulvus*, *Curvularia lunata*, *Saccharomyces cerevisiae* and *Candida boidinii*. Some of the solubility data reported by Lehmann et al. [10] does not seem reasonable, however. This may be due to failure of the presumptions or in general poor accuracy of the procedure.

Similar problems seem to be encountered by a related technique applied by Baburin et al. [11]. First, the dissolved oxygen is completely removed by stripping with nitrogen. Then the solution is aerated at a constant air flow rate recording the oxygen deficit in the exit gas due to the saturation process with a "thermo-magnetic" oxygen analyzer. As with the technique of Lehmann et al. [10] the absorbed amount of oxygen is calculated by multiplying the integral of the exit gas concentration peak and the gas flow rate. However, the oxygen exit gas concentrations range only from 20.8 % vol. down to 20.6 or 20.5 % vol. and tailing occurs. Baburin et al. [11] claim a reproducibility of approximately ±3 %. The method has been applied to some nutrition media with and without antifoam agents and culture media of lysine producing bacteria using different techniques to inhibit respiration.

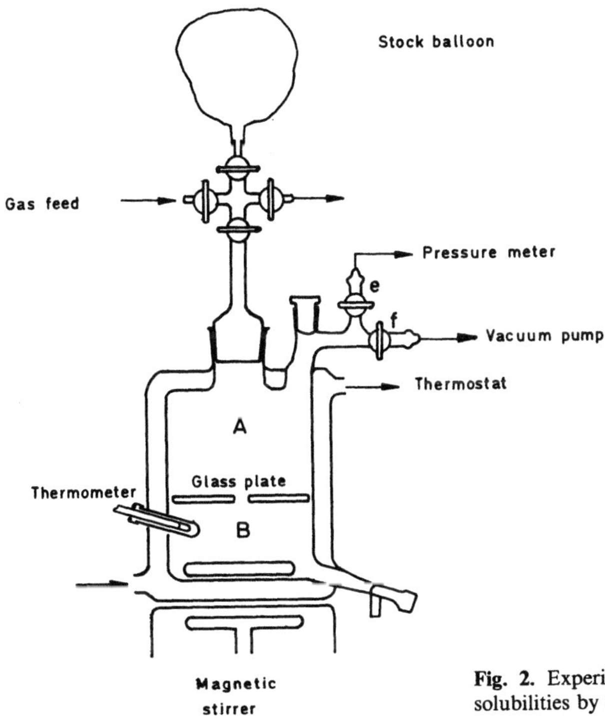

Fig. 2. Experimental setup for measuring gas solubilities by the manometric method [13]

The most accurate results still seem to be obtained by the classical saturation method. Such measurements were carried out by Popović et al. [12] applying a volumetric method and by Quicker et al. [13] applying a manometric technique to follow the equilibration process. The apparatus used by Quicker et al. [13] is sketched in Fig. 2. The difference to the stirred cell used by Popović et al. [12] is that a glass plate initially separates the gas (A) and liquid (B) sections to avoid premature absorption. The entire apparatus is housed in an air thermostat controlled to ±0.2 °C. The measuring procedure starts with degassing the solution by applying a vacuum after some water has been added to compensate for water evaporation. Degassing is terminated when the initial volume of the solution has readjusted and the temperature of the jacketed vessel is controlled to ±0.1 K. Then dry gas is introduced to the vessel from a stock balloon. After reading the initial pressure, P_0, from a micro-manometer (a mercury manometer in the case of CO_2) the magnetic stirrer is started contacting the phases by flooding the glass plate and vortex formation.

The Bunsen coefficient α is calculated from the total pressure drop $(P_0 - P_1)$ due to equilibration:

$$\alpha = \frac{V_G V_0 (P_0 - P_1)\, 101.3}{V_L RT (P_1 - P_S)} \tag{14}$$

A typical example of the pressure decrease recorded during the measurement of the oxygen solubility in a microbial culture medium is shown in Fig. 3. Since the saturation process takes only 1–4 min depending on the viscosity it usually can be well distinguished from the low linear respiration effect. Therefore Quicker et al. [13] had only to add small amounts of formalin to dampen the effect of respiration at high biomass concentrations. Excess formalin can be expected to evaporate during the degassing treatment so that no effect of added inhibitors on the results is to be expected. Thus Quicker et al. [13] determined the oxygen (and carbon dioxide) solubilities in sugar solutions, nutrition media and different culture media of *Penicillium chrysogenum*. Some unpublished results with cultures of *Saccharomyces cerevisiae*, *Chaetomium cellulolyticum*, *Hansenula polymorpha*, *Trichoderma reesei* and *Escherichia coli* are included in this paper (Sect. 7.1). An earlier paper by Popović et al. [12] also reports on oxygen solubilities in various nutrition media and in culture media of *Candida utilis*, *Saccharomyces cerevisiae*, *Aspergillus niger* and *Penicillium chrysogenum*.

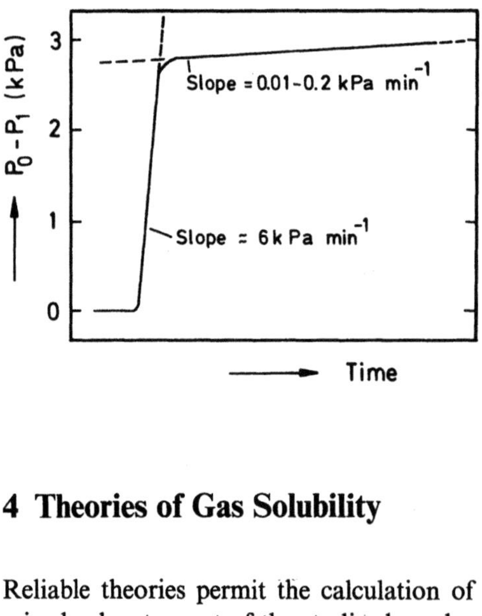

Fig. 3. Measured pressure drop as a function of time

4 Theories of Gas Solubility

Reliable theories permit the calculation of gas solubilities in pure liquids [6]. With mixed solvents most of the studies have been confined to nonaqueous systems. The theoretical access to gas solubilities in aqueous solutions of organic substances is very limited. The effect of organic solutes on the gas solubility can be rather complex, e.g., the solubility of argon in aqueous ethanol solutions has been shown [14] to run through a maximum at low ethanol concentrations and to increase again after a minimum. The peaks vanish at higher temperatures. The observations can be explained by changes in the water structure [14] but no theory permits reliable predictions. Semi-empirical correlations of nonpolar gas solubilities in aqueous alcohol solutions were suggested by Tokunaga [15] based on excess quantities and by Kojima and Tochigi [16]. The deviations are smaller with the former approach but, nevertheless, the experimental data should be referred to if available (see Sect. 5.3.2).

The solubility of gases in aqueous electrolyte solutions has been the subject of numerous investigations. A reduction of the gas solubility by the salts, a salting-out effect, is observed almost exclusively. In other words, the activity coefficients of dissolved gases are usually increased by electrolytes. The salting-out effect has been reviewed by Long and McDevit [17] and Konnik [18].

For low concentrations of electrolytes (c_{el}) and gases $j(c_j)$ the interactions can be neglected and the logarithm of the gas activity coefficient can be expressed as a linear function of both concentrations [17]:

$$\log f_j = k_s c_{el} + k_j c_j \, . \tag{15}$$

The gas solute activity in an electrolyte solution has the same value as in water if both are in equilibrium with the same gas phase. Therefore the following relations hold:

$$f_j c_j = f_{j0} c_{j0} \tag{16}$$

$$\log f_j = \log \frac{c_{j0}}{c_j} + \log f_{j0} \, . \tag{17}$$

Introducing the respective expression for the logarithms of the activities (Eq. (15)) results in

$$k_s c_{el} + k_j c_j = \log \frac{c_{j0}}{c_j} + k_j c_{j0} \tag{18}$$

and by rearrangement is obtained

$$\log \frac{c_{j0}}{c_j} = k_s c_{el} + k_j (c_j - c_{j0}) \, . \tag{19}$$

If k_j or $(c_j - c_{j0})$ are small Eq. (19) transforms to the wellknown empirical Sechenov [19] equation:

$$\log \frac{c_{j0}}{c_j} = K_s c_{el} \, . \tag{20}$$

The salting-out effects can usually be fairly well described by the so-called Sechenov constants, K_s, which are specific with respect to gas, temperature and salt.

Several theories predict salting-out coefficients. The fundamental ideas can be classified into the (1) hydration, (2) electrostatic or dispersion, (3) internal pressure and (4) scaled-particle approach [6, 17, 18].

The hydration theories are based on the idea that the decrease in gas solubility is due to hydration of the ions which reduces the volume of water available to the solute gas. This concept dominates in the early literature but cannot explain salting-in effects nor the dependency on the type of gas.

The electrostatic approach relates to the work of Debye and McAuley [20]. The electrostatic theory allows for the effect of the dissolved gas on the dielectric constant of the solution and correctly predicts K to increase with increasing charges and decreasing radius of the ions. The basic theory which assumed the solvent to be a continuous medium has been subject to considerable improvement [17,21-24].

The internal pressure concept introduced by McDevit and Long [25] considers an effective pressure related to volume and compressibility changes due to electrolyte-solvent interaction. Although many of the effects are correctly described, its applicability suffers from lack of model parameters of sufficient accuracy as do most theoretical approaches.

A promising concept is based on the scaled-particle theory which considers the dissolution process to consist of the creation of a cavity in the solvent and the introduction of a gas molecule which then interacts with the solvent. The theory, well established in case of pure solvents, has recently been extended to electrolyte solutions by Shoor and Gubbins [26] and Masterson and Lee [27].

Konnik [18] analysed several theories with regard to the correlation of theoretical solubility predictions and experimental data. None of the tested theoretical equations achieved the accuracy of a merely semitheoretical approach suggested by van Krevelen and Hoftijzer [28] which relates $\log (c_{j0}/c_j)$ to the ionic strength. This model and a recent modification which is recommended for the estimation of salting-out effects are thoroughly discussed in Sect. 5.3.1.

5 Parameters Affecting Gas Solubilities in Microbial Culture Media

5.1 Pressure

As far as biochemical engineering is concerned total pressure is usually not far from atmospheric pressure. Under this condition, proportionality of solubility and partial pressures (Henry's law) can be assumed even in the case of CO_2 without introducing appreciable errors.

5.2 Temperature

Within the range of interest the gas solubilities of most gases strongly decrease with increasing temperature.

For O_2 solubility in water the Bunsen coefficients α_0 in the temperature range of 0–50 °C reported by 11 investigators have been compared by Battino and Clever [3]. To the mean values at 5 degree intervals (Table 1) Hitchman [1] fitted an empirical correlation as follows:

$$\alpha_0 = a + bt + ct^2 + dt^3 + et^4 \qquad (21)$$

(t in °C).

Table 1. Bunsen coefficients α_0 for the O_2 and CO_2 solubilities in water (mean of literature data)

t, °C	0	5	10	15	20	25
$10^2\alpha_0(O_2)$	4.901	4.294	3.811	3.417	3.101	2.843
$\alpha_0(CO_2)$	1.720	1.422	1.193	1.008	0.872	0.754

	30	35	40	45	50	Ref.
$10^2\alpha_0(O_2)$	2.630	2.463	2.316	—	2.085	1)
$\alpha_0(CO_2)$	0.665	0.593	0.527	0.477	0.437	29–43)

The optimized coefficients of the power series according to Hitchman [1] are listed in Table 2.

To develop a corresponding correlation for the solubility of CO_2 in water within the same temperature range (0–50 °C) the data reported in Refs. (29–43) are used. In Table 1, again the mean of the α_0 values at 5 degree intervals and in Table 2 the coefficients of the power series (Eq. (21)) are listed. Fig. 4 gives a graphical representation of both correlations and the fit of the data (Table 1) they are based on.

The Bunsen coefficients α_0 calculated from Eq. (21) or obtained from Fig. 4 can readily be converted into other solubility measures by the conversion formulas given in Chapter 2, e.g., c (mg l^{-1}) is given by

$$c(\text{mg}/l^{-1}) = \frac{10^6 M_G p}{V_0\, 101.3}\,\alpha \tag{22}$$

where p (kPa) is the partial pressure of the gas, M_G (g mole^{-1}) is the molecular weight and V_0 ($\text{cm}^3\ \text{mole}^{-1}$) is the gas molar volume at 0 °C, 101.3 kPa. The values of V_0 are $V_0(O_2) = 22.395\ \text{cm}^3\ \text{mole}^{-1}$ $V_0(CO_2) = 22.258\ \text{cm}^3\ \text{mole}^{-1}$ [44]. E.g., for the O_2 concentration in equilibrium with air saturated with water vapor follows from Eq. (22)

$$c(\text{mg } O_2 l^{-1}) = \frac{10^6 \times 32 \times 0.2094\,(P_{tot} - P_s)}{22.395 \times 101.3}\,\alpha = 2.954\,(P_{tot} - P_s)\,\alpha \tag{23}$$

Table 2. Coefficients of power series (Eq. (21)), for O_2 and CO_2 solubilities as a function of the temperature (0–50 °C)

Coefficient Eq. (21)	O_2	CO_2
a	4.900×10^{-2}	1.720
b	-1.335×10^{-3}	-6.689×10^{-2}
c	2.759×10^{-5}	1.618×10^{-3}
d	-3.235×10^{-7}	-2.284×10^{-5}
e	1.614×10^{-9}	1.394×10^{-7}

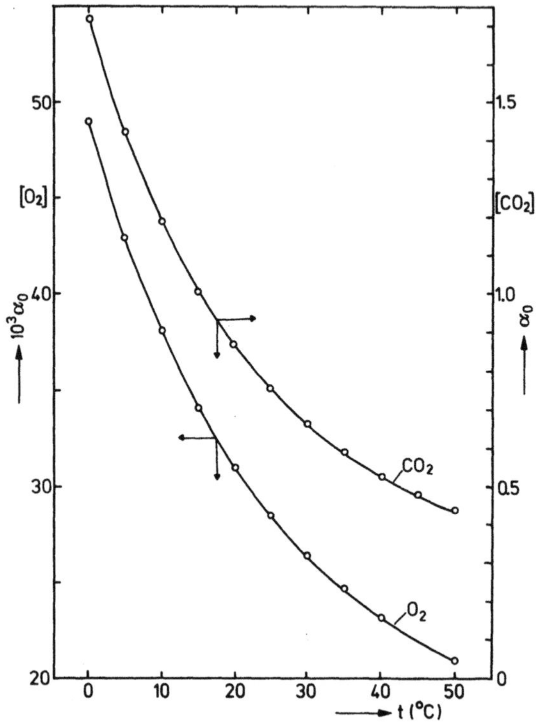

Fig. 4. Bunsen coefficients of O_2 and CO_2 in water

A comprehensive review on the solubility of gases in water has recently been given by Wilhelm et al. [45]. The temperature dependencies of the equilibrium gas mole fractions x at 101.3 kPa (1 atm) gas partial pressure were correlated by the following equation:

$$R \ln x = A + \frac{B}{T} + C \ln T + DT . \tag{24}$$

(T in K) .

For some gases of interest the coefficients according to Wilhelm et al. [45] are listed in Table 3.

5.3 Composition

As already mentioned in Sect. 4 electrolytes as well as organic substances in aqueous solution usually decrease the solubility of gases as compared to pure water. Only in a few cases, e.g., with short-chain alcohols a solubility increase may occur. This chapter is intended to supply information on the individual effects of substances which are possible components of microbial culture media.

Table 3. Coefficients of the Wilhelm et al. [45] correlations, Eq. (24), for the temperature dependency of gas solubilities in water

Gas	Temperature range, K	A	B	C	D
H_2	274–339	−357.802	13897.5	52.2871	−0.0298936
N_2	273–346	−327.850	16757.6	42.8400	0.0167645
O_2	274–348	−286.942	15450.6	36.5593	0.0187662
O_3	277–293	− 29.7374	3905.44		
CO	273–353	−341.325	16487.3	46.3757	
CO_2	273–353	−317.658	17371.2	43.0607	−0.00219107
CH_4	275–353	−365.183	18106.7	49.7554	−0.000285033
C_2H_6	275–353	−533.392	26565.0	74.6240	−0.00457313
C_2H_4	287–346	−303.888	15817.6	40.7591	
C_3H_8	273–347	−628.866	31638.4	88.0808	
n-C_4H_{10}	273–349	−639.209	32785.7	89.1483	
H_2S	273–333	−297.158	16347.7	40.2024	0.00257153

5.3.1 Single and Mixed Electrolytes

As discussed in Chapter 4, the salting-out of gases can usually be fairly well described by the empirical Sechenov equation:

$$\log \frac{c_0}{c} = K_s c_{el} \,. \tag{25}$$

However, in a plot of $\left(\log \frac{c_0}{c}\right)$ vs. c_{el} a deviation from linearity occurs at high electrolyte concentrations, the solubility predictions of Eq. (25) becoming too low. The critical electrolyte concentration beyond which Eq. (25) is no longer valid may be higher than 7 mole l^{-1} (e.g. with $NaNO_3$) but can also be less than 1 mole l^{-1}. Particularly with acids that show a concentration dependent degree of dissociation deviations from Eq. (25) can be observed at low concentrations. Sechenov constants, K_s, calculated only from gas solubilities measured at high electrolyte concentrations could therefore be too small. The limitation to moderate electrolyte concentrations should also be kept in mind when using the following semitheoretical models suggested for calculation of K_s.

Van Krevelen and Hoftijzer [28] correlated $\left(\log \frac{c_0}{c}\right)$ with respect to the ionic strength instead of merely the electrolyte concentration:

$$\log \frac{c_0}{c} = hI \tag{26}$$

where

$$h = h_+ + h_- + h_G \tag{27}$$

and

$$I = \frac{1}{2} \sum_i c_i z_i^2 \tag{28}$$

The Sechenov constants are then given by

$$K_s = h \frac{1}{2} \sum_i x_i z_i^2 \tag{29}$$

where x_i is the number of ions of type i in the salt as

$$c_i = x_i c_{el} \,. \tag{30}$$

h_+, h_- and h_G are empirical parameters specific for the cations, anions and the gas, respectively, of which only h_G is assumed to depend on the temperature.

More comprehensive parameter sets than the ones given by van Krevelen and Hoftijzer [28] were evaluated by Danckwerts [46] and Onda et al. [47,48]. That is, only the solubility parameters used differ from the original model, Eq. (26), i.e., Danckwerts uses the ratio of Henry's constant $\left(\log \frac{H}{H_0}\right)$ and Onda et al. [47] correlate the ratio of the Bunsen coefficients $\left(\log \frac{\alpha_0}{\alpha}\right)$. However, except for very high solubilities these differences are artificial because the different solubility parameters are virtually proportional to each other (see Chapter 2) and therefore

$$\log \frac{c_0}{c} = \log \frac{H}{H_0} = \log \frac{\alpha_0}{\alpha} \,. \tag{31}$$

The most comprehensive parameter set was given by Onda et al. [47,48]. Values of h_G for different gases and temperatures according to these authors are listed in Table 4. The ion specific parameters (h_+, h_-) are listed in Table 5.

For the calculation of gas solubilities in mixed electrolyte solutions both

Table 4. Selected parameters h_G (in $l\ mole^{-1}$) for the van Krevelen-Hoftijzer model (Eqs. (26)–(28)) after Onda et al. [47]

Gas	H_2	N_2	O_2	CO_2	C_2H_4	H_2S
t = 0 °C			−0.1653	−0.2110		
5 °C	−0.2106					
10 °C	−0.2170					
15 °C	−0.2197		−0.1786	−0.2222	−0.2003	
20 °C	−0.2132		−0.1771			
25 °C	−0.2115	−0.1904	−0.1892	−0.2277	−0.1951	−0.2551
40 °C				−0.2327		

Table 5. Ion specific parameters (h_+, h_-) for the van Krevelen-Hoftijzer model (Eqs. (26)–(28)) after Onda et al.[47,48]

Cation	h_+, l mole^{-1}	Anion	h_-, l mole^{-1}
H^+	−0.1110	Cl^-	0.3416
Li^+	−0.0416	Br^-	0.3310
Na^+	−0.0183	J^-	0.3124
K^+	−0.0362	OH^-	0.3875
Rb^+	−0.0449	NO_3^-	0.3230
Cs^+	−0.0584	CNS^-	0.2612
NH_4^+	−0.0737	HS^-	0.3718
		HSO_3^-	0.3869
Mg^{2+}	−0.0568	HCO_3^-	0.4286
Ca^{2+}	−0.0547		
Ba^{2+}	−0.0473	SO_4^{2-}	0.3446
Sr^{2+}	−0.0445	SO_3^{2-}	0.3275
Mn^{2+}	−0.0625	CO_3^{2-}	0.3754
Fe^{2+}	−0.0602		
Co^{2+}	−0.0534	PO_4^{3-}	0.3265
Ni^{2+}	−0.0520		
Zn^{2+}	−0.0590		
Cd^{2+}	−0.0062		
Al^{3+}	−0.0726		
Cr^{3+}	−0.0986		

Danckwerts[46] and Onda et al.[48] employed a "log-additivity" of the individual salting-out effects, i.e.

$$\log \frac{\alpha_0}{\alpha} = K_1 c_{el,1} + K_2 c_{el,2} + \dots \tag{32}$$

In terms of the van Krevelen-Hoftijzer model, Eq. (26) the following model has been suggested[46,48]

$$\log \frac{\alpha_0}{\alpha} = h_1 I_1 + h_2 I_2 + \dots \tag{33}$$

where $I_1(I_2, \dots)$ is the ionic strength attributable to salt 1(2, ...) and $h_1(h_2, \dots)$ is calculated from Eq. (27).

These van Krevelen-Hoftijzer type models, Eqs. (26) and (33), have been frequently applied; however, Schumpe et al.[49] have recently pointed out that they are physically inconsistent. Since it is suggested in Eq. (26) that for single electrolyte solutions the ion specific parameters (h_+, h_-) are to be multiplied by the total ionic strength the salting-out effect of an ion would depend on the respective counter-ion. For instance, the coefficient h_{Na^+} has to be multiplied by $I = 2$ with a 2M NaCl solution and by $I = 3$ with a 1M Na_2SO_4 solution although the Na^+ concentrations are the same in both solutions. In the case of mixed electrolyte solutions the predictions of Eq. (33) therefore depend on how the ions are arranged. For instance, the prediction for a solution of 1M K_2SO_4 and 2M NaCl differs from that for 1M Na_2SO_4 and 2M KCl although both solutions contain the same amount of each ionic species. By comparison with experimental solubility data Quicker et al. [13] showed that severe errors are possible.

Therefore Schumpe et al. [49] proposed the following model

$$\log \frac{\alpha_0}{\alpha} = \sum_i H_i I_i \tag{34}$$

where

$$I_i = \frac{1}{2} c_i z_i^2 . \tag{35}$$

I_i is the ionic strength attributable to a single ion, i, and the parameter H_i is specific to the gas, the ion and the temperature. For single salt solutions the Sechenov constants are then given by

$$K_s = \frac{1}{2} \Sigma H_i x_i z_i^2 . \tag{36}$$

Table 6. Parameters H_i (of cations in $l\ mole^{-1}$) for the model suggested by Schumpe et al. [49] (Eqs. (34)–(36))

Cations	$H_i(O_2)$			$H_i(CO_2)$
	20 °C	25 °C	37 °C	25 °C
H^+	–0.771	–0.776	–0.803	–0.319
Li^+	–0.655	–0.675	–0.636	–0.178
Na^+	–0.570	–0.568	–0.577	–0.130
K^+	–0.593	–0.587	–0.578	–0.196
Rb^+		–0.618	–0.604	–0.217
Cs^+	–0.613	–0.659	–0.612	–0.243
NH_4^+		–0.704	–0.681	–0.252
NEt_4^+			–0.709	
Mg^{2+}	–0.308	–0.297	–0.321	–0.078
Ca^{2+}	–0.293	–0.309	–0.316	–0.073
Ba^{2+}		–0.291	–0.299	–0.064
Mn^{2+}		–0.324	–0.325	–0.084
Fe^{2+}				–0.078
Co^{2+}			–0.317	
Ni^{2+}	–0.302		–0.318	
Cu^{2+}	–0.312		–0.325	–0.090
Zn^{2+}	–0.295		–0.310	
Cd^{2+}	–0.316		–0.320	
Al^{3+}	–0.210		–0.221	–0.059
La^{3+}			–0.216	
Ce^{3+}			–0.216	
Fe^{3+}			–0.244	
Th^{4+}			–0.168	

Equation (34) applies not only to single electrolyte solutions but is directly applicable also to mixed electrolyte solutions containing an arbitrary number of different ionic species. The model has also been successfully extended to mixed solutions of electrolytes and organic substances which are to be discussed in Chapter 6.

H_i parameter values of the model suggested by Schumpe et al.[49] have been reported for salting-out O_2 and CO_2 at 25 °C in Refs. [13,49,50]. New versions of these parameter sets based on revised and updated literature data [34,35,43,47,49,51–62] and data from our own updated measurements in some doubtful cases are listed in Tables 6 and 7. In addition, parameter sets for salting-out O_2 at 20 °C and 37 °C were evaluated from comprehensive solubility data supplied by Lang[62]. The experimental technique applied is a desorption method with subsequent reabsorption in alkaline catechol/$Fe(NH_4)_2(SO_4)_2$ solution and photometric analysis[63]. The H_i values for salting-out O_2 at different temperatures agree rather well. In general, a slight decrease of K_s with increasing temperature is observed, however, this effect is often smaller than the scatter of the experimental data. For practical applications it is recommended to use the H_i parameter set for the nearest temperature rather than any interpolation. Furthermore, the calculation should be based on a rough chemical understanding of the actual composition of the solution, i.e., neutralisation

Table 7. Parameters H_i (of anions in l mole^{-1}) for the model suggested by Schumpe et al.[49]

Anions	$H_i(O_2)$			$H_i(CO_2)$
	20 °C	25 °C	37 °C	25 °C
F^-			0.867	
Cl^-	0.843	0.849	0.861	0.339
Br^-	0.840	0.820	0.822	0.324
J^-		0.784		0.309
OH^-	0.955	0.943	0.917	
NO_3^-	0.827	0.802	0.821	0.293
SCN^-			0.791	
BF_4^-			0.775	
ClO_4^-		0.890		
HSO_4^-	0.945	0.955	0.935	0.436
HSO_3^-				0.400
HCO_3^-		1.076	0.861	
$H_2PO_4^-$		0.997		
C_6H_5—O—Ac^-		0.755		
SO_4^{2-}	0.458	0.460	0.448	0.213
$S_2O_3^{2-}$		0.455		0.211
SO_3^{2-}				
CO_3^{2-}		0.467	0.447	
HPO_4^{2-}		0.477		
PO_4^{3-}		0.308		
$Mo_7O_{24}^{6-}$			0.155	

of acidic and basic ingredients (e.g., 1 M Na^+/OH^- + 1 M H^+/Cl^- = 1 M Na^+/Cl^-) and limited dissociation of acids, in particular (e.g., H_2SO_4 = H^+/HSO_4^- but not $2H^+/SO_4^{2-}$, KH_2PO_4 = $K^+/H_2PO_4^-$ but not K^+2H^+/PO_4^{3-}) have to be taken into consideration.

Figure 5 is a parity plot of all the Sechenov constants available from the literature for O_2 at 20, 25 and 37 °C and the values calculated from Eq. (34) by using parameters listed in Tables 6 and 7. In most cases, a striking agreement between experimental and calculated K_s values is found. The model suggested by Schumpe et al.[49] also holds in the case of mixed electrolyte solutions where "log-additivity" of the salting-out effects of the individual ions is assumed. This is in accordance with Eq. (32), but Eq. (34) avoids the inconsistency encountered with the van Krevelen-Hoftijzer approach (Eq. (33)). Nevertheless, the latter model still has to be applied to gases other than O_2 and CO_2 where no H_i parameter sets are available.

5.3.2. Organic Compounds

For many organic solutes the concentration dependency of gas solubilities can be described in an analogous manner to the salting-out effect (cf. Eq. (25)):

$$\log \frac{\alpha_0}{\alpha} = Kc_n \,. \tag{37}$$

Here c_n is the concentration of the organic solute and K is an empirical constant which corresponds to the Sechenov constant, K_s. Occasionally a better fit is obtained by a linear relation:

$$\alpha = \alpha_0(1 - mc_n) \tag{38}$$

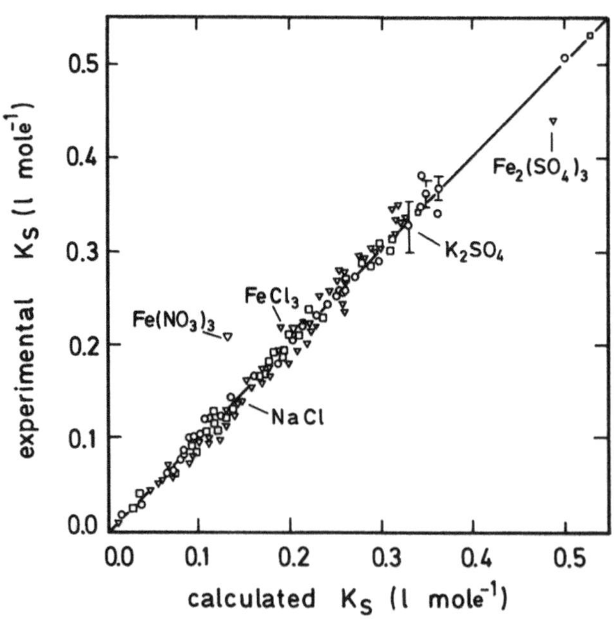

Fig. 5. Parity plot of experimental Sechenov constants for O_2 (20, 25 and 37 °C) and those calculated from Eq. (36) with H_i data listed in Tables 6 and 7

or a two-parameter model while in some cases no simple relation can be given at all. There are no uniform correlations of the individual effects of organic substances as in the case of electrolytes (Sect. 5.3.1). Therefore the discussion is confined to selected substances and the solubilities of O_2 and CO_2. For a convenient access to additional experimental data the "Solubility Data Series"[64] may be referred to; however, only a few volumes of the series have appeared so far.

The gas solubilities in small-chain alcohols are higher than in water. For alcohol-water mixtures the dependency on the alcohol concentration is not a linear function but maxima and minima may even exist at low concentrations. For instance, when studying the argon solubility in aqueous ethanol Ben-Naim and Baer[14] found at low temperatures a slight increase for small ethanol mole fractions ($x_n < 0.05$) followed by a solubility decrease beyond the solubility in water and a strong increase at higher ethanol mole fractions ($x_n > 0.15$). At higher temperatures ($t > 24$ °C) the minimum vanished and the argon solubility increased steadily though nonlinearly with the ethanol concentration.

O_2 solubilities in low-chain alcohols have also been frequently investigated but data at low alcohol concentrations is scarce and disagrees widely, particularly, in the case of ethanol. In Fig. 6, the ratios α/α_0 of the reported solubilities referred to the solubility in pure water at the same temperatures are plotted vs. the mole fraction of ethanol. While differing from the salting-out by electrolytes the ratio α/α_0 is obviously strongly temperature dependent. At low temperatures the data of both Tokunaga[65] and Shchukarev and Tolmacheva[66] suggest a reduced O_2 solubility at small deviations from α_0 at low ethanol concentrations. From the data of Tokunaga[65] for 20 and

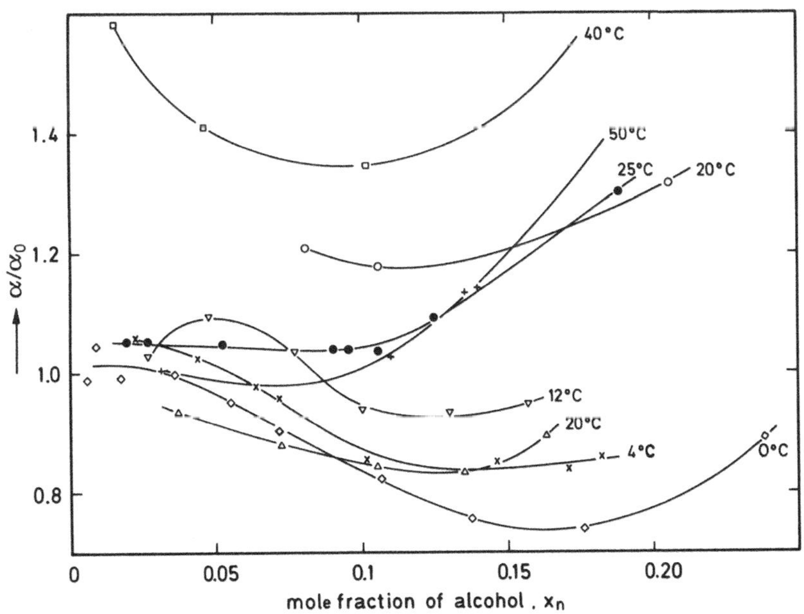

Fig. 6. Solubilities of O_2 in aqueous ethanol solutions; (□) (○) (◇) Tokunaga[65], (+) (●) (▽) (×) Shchukarev and Tolmacheva[66], (△) Lubarsch[67]

40 °C, however, a tremendous increase in the O_2 solubility would have to be expected at small x_n although other investigators [50, 66, 67] observed only a moderate increase. Also for aqueous methanol Tokunaga [65] reports relatively high values of α/α_0 whereas other investigations suggest only a slight increase [50] or even a very slight decrease [68]. By adding 1–3% vol. methanol to a mixed electrolyte solution Popović et al. [12] also observed a decrease of the O_2 solubility. For propanol the results of Tokunaga [65] on the O_2 and CO_2 solubility appear to be more consistent. Minima with $\alpha/\alpha_0 < 1$ are observed for both gases but at very low concentrations where no measurements were made there might also be some increase.

In general, data for low alcohol concentrations is scarce and not at all conclusive. There is probably some moderate increase at low concentrations of methanol and ethanol but more intense studies in this concentration range are needed to clarify this point.

The effects of sugars frequently used as a carbon source in microbial cultivations have been intensely investigated by Quicker et al. [13] with glucose, lactose and sucrose at a temperature of 25 °C. By plotting $\log \frac{\alpha_0}{\alpha}$ vs. the sugar concentrations (Figs. 7 and 8) straight lines are obtained in accordance with Eq. (37) for both O_2 and CO_2. The results of Quicker et al. [13] on the reduction of the CO_2 solubility by sucrose agree fairly well with the data of Findlay and Shen [69] and Koch [70]. For O_2 the observed solubility decrease was smaller than reported by Hikita et al. [71]. Furthermore Eq. (37) was found to be valid up to a sucrose concentration of about 200 g l^{-1} while only Eq. (38) with $m = 9.04 \times 10^{-4}$ l g^{-1} for O_2 and $m = 6.87 \times 10^{-4}$ l g^{-1} for CO_2 applied to the whole range of concentrations investigated [13]. Baburin et al. [11] suggested a coefficient of $m = 15.6 \times 10^{-4}$ l g^{-1} for the effect of sucrose on the O_2 solubility at 30 °C. This value is obviously too high. From measurements only at low concentrations Popović et al. [12] proposed a uniform coefficient of $m = 12 \times 10^{-4}$ l g^{-1} at

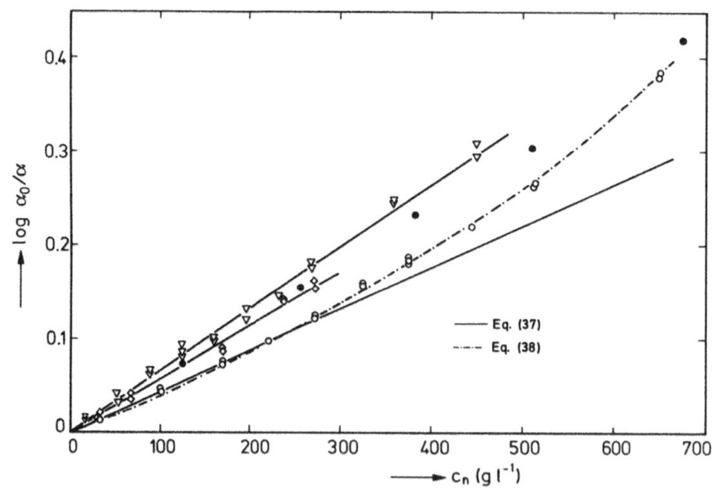

Fig. 7. O_2 solubilities at 25 °C in solutions of glucose (▽), lactose (◇), Quicker et al. [13], and sucrose (●) Hikita et al. [71], (○) Quicker et al. [13]

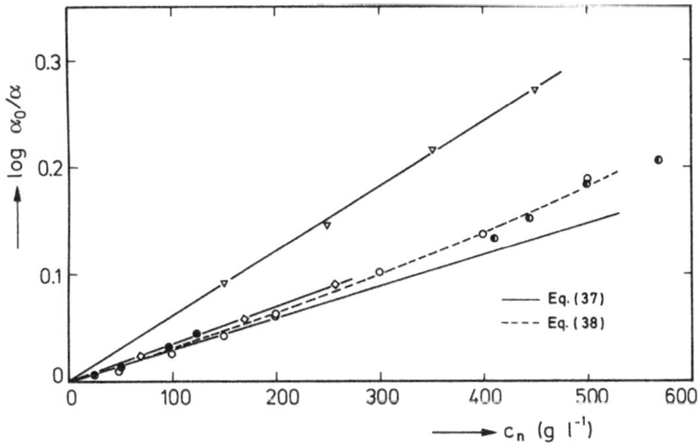

Fig. 8. CO_2 solubilities in solutions of glucose (▽), Koch[70]; lactose (◇), Quicker et al.[13], and sucrose (●), Findlay and Shen[69]; (◐), Koch[70], (○), Quicker et al.[13]

25 °C to account for the effects of not only sucrose but also of glucose and lactose. However, this turns out to be too rough an approximation if a wider concentration range is investigated (cf. Fig. 7). Parameter values K and m, respectively, for glucose, lactose and sucrose are listed in Tables 8 and 9. From these and other literature data[75-78] it can be concluded that the effects depend only slightly on the temperature but are specific for the sugar and the gas.

Quicker[73] and Quicker et al.[13] also measured O_2 solubilities in mixed solutions of different sugars and found the individual effects to be "log-additive" just as in the case of salts (cf. Eq. (32)).

$$\log \frac{\alpha_0}{\alpha} = K_1 c_{n1} + K_2 c_{n2} + \dots . \tag{39}$$

K values for various other organic substances are also listed in Tables 8 and 9. Some original data for which the application of Eq. (37) does not seem to be justified are listed in Table 10. Also with some other substances for which K values are specified the experimental data may be too scarce or scattered to strictly prove the applicability of Eq. (37) but the parameters listed at least provide a reasonable estimate of the concentration dependencies. For instance, in the case of citric acid and for glycerol, Eq. (37) is an approximation restricted to low concentrations. To fit their data on citric acid in the whole concentration range investigated, Sada et al.[74] used a two-parameter model. The K values for citric acid are smaller for CO_2 than for O_2 as observed also for sugars and electrolytes. In the case of several amino acids, Zander[72] reports rather high salting-out effects while the effects of proteins are relatively small. With albumin Baburin et al.[11] have recently reported a strong increase in the O_2 solubility which was attributed to adsorption of O_2. This finding, however, disagrees with the results of Zander[72] and Quicker[73] who observed a decrease of the O_2 solubility with increasing albumin concentration. Nevertheless,

Table 8. Parameters K (Eq. 37) for the effects of organic substances on O_2 solubilities

Substance	Concentration range $g\,l^{-1}$	Temperature °C	K $10^{-4}\,l\,g^{-1}$	Ref.
Glucose	0–450	25	6.58	13)
	0–200	37	6.78	72)
Lactose[a]	0–300	25	5.71	13)
Sucrose	0–200[b]	25	4.36	13)
	0–700	15–45	5.99	71)
	0–200	37	5.19	72)
Molasses[c]	0–240	25	4.03	12)
Dextrin	0–200	37	5.02	72)
Insulin	0–200	37	5.48	72)
Starch	0–200	37	6.35	72)
Glycogen	0–250	37	6.59	72)
Glucosamine	0–300	37	11.23	72)
Glucose-(1)-phosphate	0–200	37	11.87	72)
ATP	0–200	37	7.10	72)
ADP	0–200	37	6.35	72)
Gluconic acid	0–500	25	3.92	12)
Citric acid	0–200	25	5.09	12)
Urea	0–300	37	3.74	72)
Glycerol	0–300	37	4.07	72)
Albumin (bovine)	0– 80	25	1.60[d]	73)
	0–200	37	1.81	72)
Albumin (chicken)	0–200	37	3.23	72)
α-Globulin	0–100	37	3.05	72)
β-Globulin	0–150	37	3.72	72)
γ-Globulin	0–200	37	2.56	72)
Hemoglobin	0–250	37	−0.30	72)
Hydroxyproline	0–200	37	7.81	72)
β-Alanine	0–300	37	10.38	72)
Glycine	0–200	37	12.46	72)
Lysine	0–300	37	13.45	72)
Cysteine	0–200	37	22.82	72)
Caseinpeptone[c]	0– 60	30	4.3	73)
Meat extract[c]	0– 60	30	5.7	73)
Yeast extract[c]	0– 60	30	6.2	73)
Pharmamedia[c]	0– 80	25	1.5	13)

[a] technical grade;
[b] Eq. (38) with $m = 9.04 \times 10^{-4}\,l\,g^{-1}$ valid up to $600\,g\,l^{-1}$;
[c] measured after sterilisation;
[d] fraction V, pH = 5

adsorption on solid surfaces or macromolecules might occur in some cases and increase the absorption capacity. This will be discussed in the following chapter.

5.3.3. Adsorption Effects

In albumin solutions increased overall absorption capacities for gaseous hydrocarbons result from interactions of the hydrocarbons with the apolar groups of the protein

Table 9. Parameters K (Eq. 37) for the effects of organic substances on CO_2 solubilities

Substance	Concentration range $g\,l^{-1}$	Temperature °C	K $10^{-4}\,l\,g^{-1}$	Ref.
Glucose	0–300	15	6.63	[75]
	0–450	25	6.07	[13]
Lactose	0–250	25	3.48	[13]
Sucrose	0–200[a]	25	2.94[b]	[13]
Citric acid	0–200	25	2.68	[74]

[a] Eq. (38) with $m = 6.87\,l\,g^{-1}$ valid up to $500\,g\,l^{-1}$;
[b] In close agreement with the data of Findlay and Shen [59] and Koch [70] for 25 °C and Usher [77] for 20 °C.

Table 10. Effects of organic substances on O_2 solubilities

Substance	Concentration $g\,l^{-1}$	Temperature °C	$\frac{\alpha}{\alpha_0}$	Ref.
Xanthan	1	25	0.9889	[12]
	2	25	0.9897	
Pullulan	1	25	1.0016	[12]
	5	25	0.9818	
	10	25	0.9810	
Penicillin G	18	25	0.94	[12]

which are also affected by the pH of the solution [79–82]. From these effects the enthalpy and entropy changes for the hydrocarbons are available and have been used to draw conclusions with respect to the protein structure. Also in micellar solutions, e.g., of sodium dodecylsulfate, increased absorption capacities for hydrocarbons are to be found [83, 84]. In the case of O_2 increased solubilities have been measured in the presence of antifoam agents. Baburin et al. [11] reported on an almost fourfold increase by adding only 4% vol. of propinole (oxypropylene and propylene glycol) and an almost threefold increase by adding 2% vol. of sunflower oil. These effects have been explained as an adsorption of O_2 on the oil interphase since a release of O_2 was observed to follow the coalescence of oil drops. Surprisingly, Baburin et al. [11] report no such effect to occur in a distilled water-oil mixture. Popović et al. [12], on the other hand, observed no significant solubility changes after addition of an antifoam agent.

The results of Zander [72] and Quicker [73] presented in the previous chapter indicate that adsorption on albumin does not take place in the case of O_2. On the other hand, Baburin et al. [11] reported a drastic increase in the O_2 solubility caused by albumin. This effect was then used to explain increased O_2 solubilities measured in culture media if the cells were partially lysed thus setting free proteins. The intact cells themselves, on the other hand, are suspected of reducing the oxygen solubility by an effect of the membrane potential on the structure of the solution near the cells. Their observation of an increase in the O_2 solubility by dilution of a medium with intact cells is, however, not conclusive since this is to be expected in any case due to dilution of other salting-out ingredients, e.g. dissolved electrolytes. Furthermore, Quicker et al. [13] when studying the O_2 solubility in culture media of *Penicillium chrysogenum* found no significant differences no matter whether the solubilities were measured in the presence or absence of cells. From the measurements of Popović et al. [12] no conclusions can be drawn with respect to the effect of the cells since it is not clear from their paper whether they measured in the presence or absence of biomass.

So far the few results reported in the literature pertaining to the effects of biomass on the O_2 solubility are not consistent so that this point needs to be further investigated. However, any adsorption effects could be expected to increase only the overall absorption capacity but not the level of dissolved oxygen affecting the respiration and the driving force of gas/liquid mass transfer [11].

6 Predictions of Solubilities in Media

The main components of most media are sugars and salts. To predict the gas solubility for a multitude of media it is therefore important to be able to describe the salting-out effect in solutions of sugar/salt mixtures or, more generally speaking, in electrolyte/nonelectrolyte mixtures.

In Sect. 5.3 we were able to demonstrate that the gas solubility in solutions of salt mixtures as well as the solubility in solutions of sugar mixtures can be computed in a manner analogous to the Sechenov equation. The salting-out effects of the single sugars were found to be log-additive with respect to the overall solubility reduction of the mixture. For oxygen this log-additivity behavior has already been demonstrated by Schumpe and Deckwer [50] for mixed solutions of electrolytes and nonelectrolytes (methanol, ethanol, propanol, glycerol). Therefore the above gas solubility model was extended [13] to predict oxygen solubilities in solutions containing salts as well as nonelectrolytes:

$$\log (\alpha_0/\alpha) = \sum_i H_i I_i + \sum_j K_j c_{n,j} \tag{40}$$

or in more general terms:

$$\log (\alpha_0/\alpha) = \sum_l K_l C_l = \sum_l \log (\alpha_0/\alpha_l) \tag{41}$$

where $\sum_l \log (\alpha_0/\alpha_l)$ means the contributions of the respective solutes (electrolytes and nonelectrolytes) which log-additively form the overall salting-out effect, i.e., $\log (\alpha_0/\alpha)$.

Quicker et al. [13] determined the oxygen and carbon dioxide solubilities in solutions containing sugars in addition to salts in equimolar concentrations. Figure 9 shows the results as a plot of $\log (\alpha_0/\alpha)$ vs. the equimolar concentration of the respective components. The lines are predictions of Eq. (40); they are in striking agreement with the experimental data.

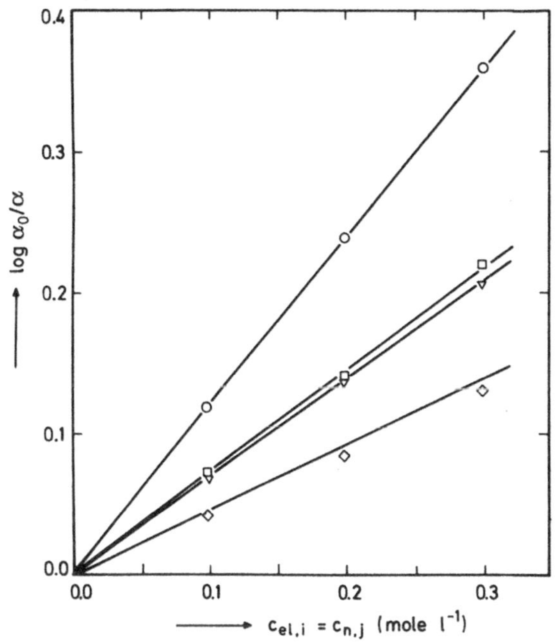

Fig. 9. O_2 solubilities at 25 °C in mixed solutions of sugars and salts (each component present at the same concentration): (○) sucrose + K_2SO_4 + K_2HPO_4 + $Na_2S_2O_3$, (□) sucrose + glucose + $CaCl_2$ + $MgCl_2$ (▽) glucose + $Na_2S_2O_3$ + $CaCl_2$, (◇) sucrose + glucose + lactose, — predicted from Eq. (40) with parameters given in Tables 6 to 8

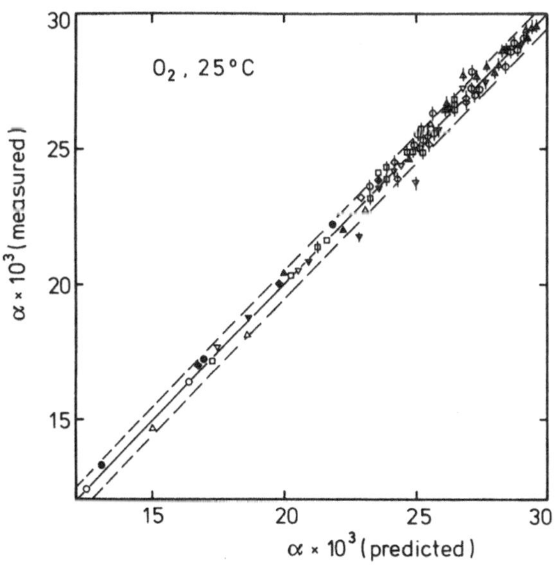

Fig. 10. Comparison of measured solubilities and predictions for mixed solutions of salts and sugars and alcohols (after Quicker et al. [13])

A comparison of calculated and measured oxygen solubilities [13] at 25 °C in solutions of both salts and organic substances (carbohydrates and alcohols) is given in Fig. 10. The discrepancy between predicted and measured α values lies within α ± 0.0005. Popović et al. [12] also investigated the oxygen solubility in sugar-electrolyte mixtures. To predict the solubility they also applied a log-additive relation like Eq. (41). Agreement was found within $\pm 2\%$ for experimental and predicted values.

Obviously, oxygen solubilities in mixed solutions of electrolytes and sugars or alcohols are well described by Eqs. (40) and (41), respectively. It can be expected that the log-additive approach is valid for other organic substances and other gases too. Therefore the model was applied to calculate oxygen solubilities in different nutrition media the composition of which is given in Table 11. Additional amounts of trace ingredients in the media are not listed in Table 11. Their concentration is too low to affect the solubility and these trace ingredients as well as solids ($CaCO_3$, Avicel) are not taken into account when calculating the oxygen solubility.

Samples of the nutrition media were taken after sterilization (at 121 °C and 120–140 kPa for 20 min) but prior to inoculation. A comparison of the measured oxygen solubilities in the media given in Table 11 with calculations according to Eq. (41) is shown in Fig. 11. This figure also presents experimental data and estimates in different nutrition media for *Penicillium chrysogenum* [13] the composition of which is listed in Table 12. The discrepancy between predicted and experimental Bunsen coefficients is less than 2% with the exception of the nutrition media XII and XIII. An explanation for the poorer agreement (media XII, XIII) may be that the salting-out constants, K, for the solutes casein peptone, yeast and meat extract are estimated from oxygen solubility data at two solute concentrations only.

From the above results it can be concluded that the relationship (41) is suitable to predict the oxygen solubility in nutrition media as far as the individual salting-out constants of the components are known. Obviously, this holds also for nutrition media which mainly consist of albumens (peptones, free amino acids) such as medium XII.

Table 11. Composition of media used in different growth experiments

Component	VIII $g\,l^{-1}$	IX $g\,l^{-1}$	X $g\,l^{-1}$	XI $g\,l^{-1}$	XII $g\,l^{-1}$	XIII $g\,l^{-1}$
Glucose	8	30			20	30
Avicel[a]			5	10		
K_2HPO_4	0.76					
KH_2PO_4	3		2			
$(NH_4)_2HPO_4$		1.8				
$NH_4H_2PO_4$				2.4		
$(NH_4)_2SO_4$	5	6	3			
$MgSO_4$	0.24	0.22	0.15	0.15		
K_2SO_4				0.3		
$CaCl_2$		0.32	0.30	0.27		
KCl	0.1	0.87				
NaCl	0.1				5	7.5
Yeast extract			1		20	30
Casein peptone					20	30
Meat extract					10	15

[a] cellulose, 90 μm, Serva Chemicals

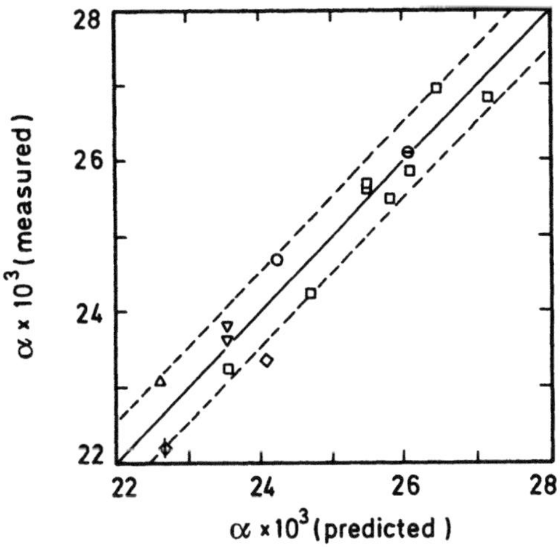

Fig. 11. Oxygen Bunsen coefficients (at various temperatures, see Table 14) of nutrition media — measured values and predictions of Eq. (41), I—VII: □, VIII: △, IX: ○, X: ▽, XI: ⊖, XII: ◇, XIII: ⏀

Table 12. Composition of nutrition media used for growth of *Penicillium chrysogenum*

Component	I $g\,l^{-1}$	II $g\,l^{-1}$	III $g\,l^{-1}$	IV $g\,l^{-1}$	V $g\,l^{-1}$	VI $g\,l^{-1}$	VII[a] $g\,l^{-1}$
Lactose	95	66.5	47.5	39.9	33.25	25.55	20.19
Sucrose	20	14	10	8.5	7	5.4	
Pharmamedia	40	28	20	17	14	10.8	12.5
$Na_2S_2O_3$	4.78	4.78	4.78	4.78	4.78	4.78	2.39
$CaCO_3$	7.5	5.3	3.8	3.2	2.6	2	
K-Phenoxyacetate	6	6	6	6	6	6	3

[a] additionally contains 0.09 $g\,l^{-1}$ KH_2PO_4 and Na_2HPO_4

A sample calculation for the oxygen solubility in medium IX is given below.

The salting-out parameter, K, for glucose is taken from Table 8 and converted from $l\,g^{-1}$ to $l\,mole^{-1}$. K values of the respective salts are computed by means of Eq. (36), the H_i values for O_2 are taken from Tables 6 and 7 using the set for $t = 25$ °C. For instance, in the case of $(NH_4)_2HPO_4$ follows:

$$K = 0.5 \sum_i H_i x_i z_i^2 = 0.5\,[-0.704 \times 2 \times 1^2 + 0.477 \times 1 \times 2^2]$$

$$K = 0.250\ (l\,mole^{-1})\,.$$

The Bunsen coefficient α of the nutrition medium is then estimated by means of Eq. (14) introducing the appropriate values from Table 13.

$$\begin{aligned}\log(\alpha_0/\alpha) &= \sum_l K_l c_l \\ &= 0.167 \times 0.119 + 0.014 \times 0.250 + 0.045 \times 0.216 \\ &\quad + 0.002 \times 0.326 + 0.003 \times 0.231 + 0.012 \times 0.131 \\ &= 0.03601\end{aligned}$$

Table 13. Sample calculation of the oxygen solubility in medium IX, t = 30 °C

Component	($g\,l^{-1}$)	($mole\,l^{-1}$)	K ($l\,mole^{-1}$)
Glucose	30	0.167	0.119
$(NH_4)_2HPO_4$	1.8	0.014	0.250
$(NH_4)_2SO_4$	6	0.045	0.216
$MgSO_4$	0.22	0.002	0.326
$CaCl_2$	0.32	0.003	0.231
KCl	0.87	0.012	0.131

The Bunsen coefficient α_0 for oxygen in water at t = 30 °C is computed by means of Eq. (21) and values from Table 1. Thus the following relationships are obtained:

$$\alpha = \alpha_0 10^{-0.03601}$$

$$\alpha = 0.02635 \times 10^{-0.03601}$$

$$\alpha = 0.02425$$

7 Estimation of Solubilities during Actual Bioreaction

The only difference between the nutrition medium discussed in Sect. 6 and the actual reaction mixture at the start of a batch lies in the microorganisms added. The growth of the microorganisms changes the composition of the medium and hence the gas solubility. Alteration of medium composition is due to conversion of C sources (e.g. sugars) and nutritional salts (N source) into cell mass and due to production of metabolic products by the biomass. Additional changes result from the addition of acid or alkali for controlling the pH and from feeding nutrients during fermentation.

7.1 Direct Predictive Method

The correlation, i.e. Eq. (41), developed for the estimation of oxygen solubilities in nutrition media can also be applied to predict the solubility during the performance of bioprocesses. But it is necessary to have information on the gross changes of the main components in the broth during a batch process.

Normally, alterations in the substrate concentration (e.g. sugar) as well as added amounts of bases and acids for a constant pH are known. In fermentations carried out with continuous feeding of salts (e.g. ammonia salts as nitrogen source) the increase in the respective salt concentration can be considered. If a certain substance is produced during a process there is usually an analytical control of this product. Therefore a calculation of the product influence on the oxygen solubility is possible.

The calculation of the oxygen solubility during a batch corresponds to the calculation for a nutrition medium (see example Sect. 6). At the start of a process the

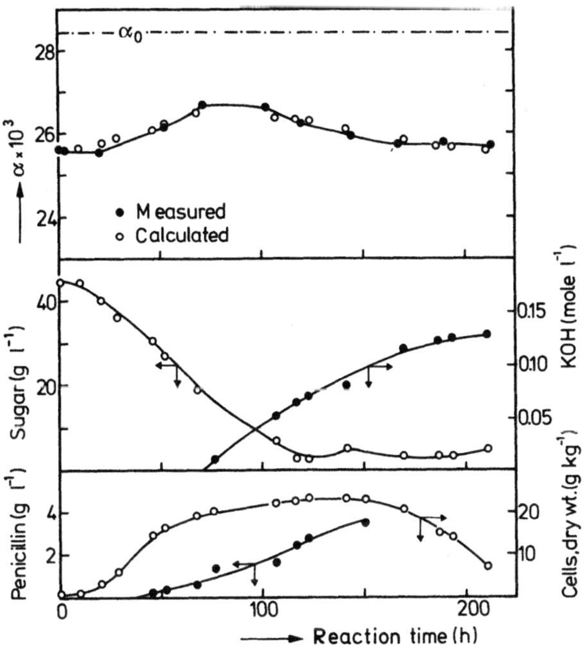

Fig. 12. Oxygen solubilities in growth of *Penicillium chrysogenum* at 25 °C

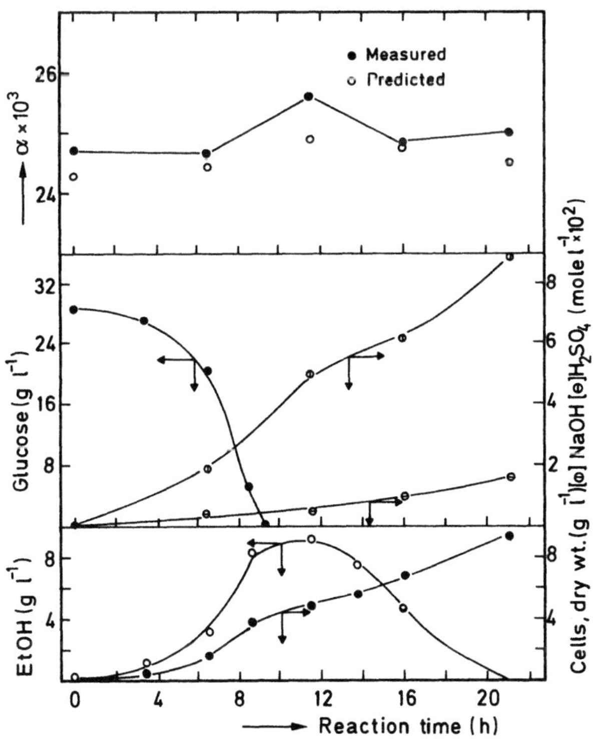

Fig. 13. Oxygen solubilities in growth of *Saccharomyces cerevisiae* at 30 °C

gas solubility in the reaction mixture is, of course, equal to that in the nutrition medium provided the addition of the inoculum does not change the composition of the medium significantly (e.g. by dilution).

Oxygen solubilities during growth of *Penicillium chrysogenum* (medium V) [13] and *Saccharomyces cerevisiae* (medium IX) are given in Figs. 12 and 13 together with additional information. The Bunsen coefficients, α, for oxygen during the course of the reactions were calculated from the changes in carbohydrate concentration and the addition of acids and bases to control the pH starting from the α-values of the respective nutrition media. The amounts of ethanol (Fig. 13) were not taken into consideration due to the lack of reliable solubility data for alcohols in the low concentration range. There probably is a small salting-in effect [65, 66].

For details on the calculations for *Penicillium chrysogenum* see Quicker et al. [13]. The composition and concentrations of the other solutes in the broth certainly change during the course of the fermentation but the salting-out effect due to this group of compounds is assumed to be constant. The good agreement between computed and measured Bunsen coefficients in Figs. 12 and 13 confirms this assumption.

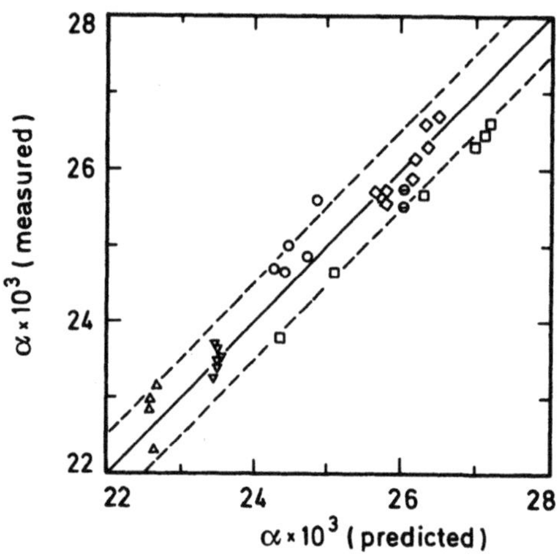

Fig. 14. Comparison of experimental and predicted O_2 solubilities in reaction mixtures *Saccharomyces cerevisiae*: ○; *Trichoderma reesei*: ⊖; *Chaetomium cellulolyticum*: ▽; *Hansenula polymorpha*: △; *Penicillium chrysogenum*: □, ◇ nutrition media VIII and V

Table 14. Experimental conditions for growth experiments

Microorganism	V_R	pH	t °C	Final cell dry weight (g l^{-1})	Nutrition medium
Penicillium chrysogenum	60–80	6.2	25	10–50	I–VII
Hansenula polymorpha	20	5.0	38	5	VIII
Saccharomyces cerevisiae	20	5.0	30	9.5	IX
Chaetomium cellulolyticum	20	5.0	37	1	X
Trichoderma reesei	2.5	5.0	30	(3)[a]	XI
Escherichia coli	45	6.8	28	12–16[b]	XII, XIII

[a] continuous cultivation
[b] sediment

Fig. 14 shows a comparison of calculated and measured values of oxygen solubilities in other media for *Penicillium chrysogenum* [13], *Trichoderma reesei*, *Chaetomium cellulolyticum* and *Hansenula polymorpha*. Agreement was found within ±2% in most cases. The experimental conditions for the fermentations are given in Table 14. Except for *Trichoderma reesei* all processes were carried out batchwise. The respective nutrition media are listed in Table 11 and 12.

7.2 Indirect Predictive Method

An interesting technique to follow O_2 solubilities in microbial culture media was proposed by Popović et al. [12]. The basis of this method is the assumption that oxygen solubility is decisively influenced by the salts present in the culture media. The concentration of these salts can be easily measured by electrical conductivity. To relate electrical conductivity and O_2 solubility Popović et al. [12] have suggested the following polynomial

$$\log\left(\frac{\alpha_0}{\alpha_{el}}\right) = a_0 + a_1 x' + a_2 x'^2 \tag{42}$$

where x' is the conductivity in Ω^{-1} cm^{-1}. a_0, a_1 and a_2 are empirical constants which have to be determined by independent conductivity and solubility measurements. Values of these coefficients (a_0, a_1, a_2) for single salts are given in Table 15. The fact that the values of the coefficients depend on other ions present in the solution needs to be taken into consideration. Therefore, in the case of mixtures of electrolytes the coefficients of Eq. (42) must be determined separately. In addition, the coefficients lose their identity if the ratios of the concentrations of certain ions in the mixture vary during fermentation.

Fig. 15 shows the salting-out of various salts as a function of the electrical conductivity. It can be seen that these salts differ significantly in their action.

In order to describe the entire influence of all the components present in a culture medium on O_2 solubilities Popović et al. [12] also used a log-additive approach.

$$\log\frac{\alpha_0}{\alpha} = \log\frac{\alpha_0}{\alpha_{el}} + \sum_j \log\frac{\alpha_0}{\alpha_j} \tag{43}$$

Table 15. Coefficients of Eq. (42) for the temperature range of 20 to 35 °C (Popović et al. [12])

Salts	Concentration mole l^{-1}	a_0	a_1	a_2
$(NH_4)_2SO_4$	0.01–0.3	0.0112	0.981	6.92
$NaNO_3$	0.05–0.3	−0.0037	2.748	−29.30
K_2HPO_4	0.05–0.3	−0.0033	3.055	−12.42
KH_2PO_4	0.05–0.5	0.0044	3.761	70.18
$MgSO_4 \cdot 7\,H_2O$	0.01–0.5	−0.0020	2.931	75.02
$(NH_4)_2HPO_4$	0.01–0.3	0.0083	1.244	67.73
$(NH_4)H_2PO_4$	0.01–0.8	0.00001	2.767	22.62
Sodium citrate	0.01–0.3	0.0072	1.810	62.66
Ammonium citrate	0.004–0.088	−0.0034	1.295	30.27
Molasses	up to 300 g l^{-1}	0.0025	1.315	—

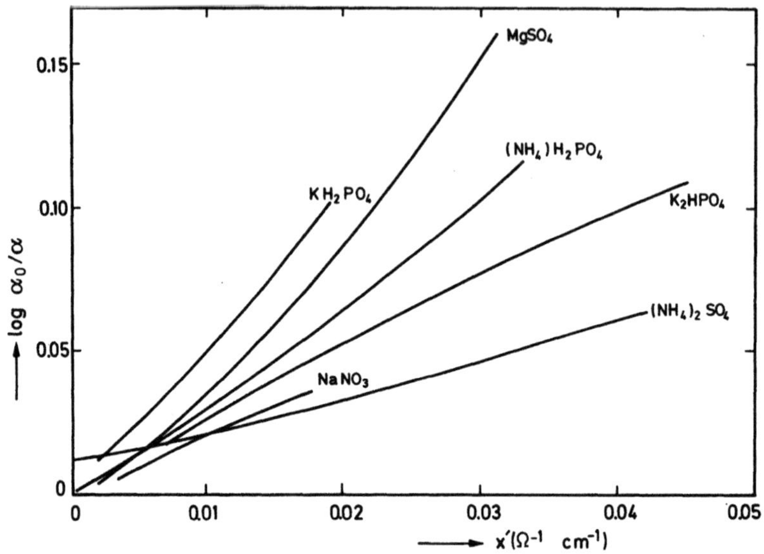

Fig. 15. Salting-out effect of some salts as a function of electrical conductivity calculated from Eq. (42) with data of Table 15

Under consideration of Eqs. (38) and (42) it follows that

$$\log \frac{\alpha_0(1 - mc_n)}{\alpha} = a_0 + a_1x + a_2x^2 \tag{44}$$

This equation can be used to calculate the oxygen solubility in the broth on the basis of measured conductivities and sugar concentrations. Popović et al. report that for growth of *Candida utilis*, *Candida boidinii*, *Saccharomyces cerevisiae*, *Aspergillus niger* and *Penicillium chrysogenum* the measured and calculated oxygen solubilities agree within $\pm 2\%$.

It is not clear in how far the gas solubility varies due to changes in conductivity as the conductivities during fermentations are not given in the paper.

During growth of *Saccharomyces cerevisiae*, molasses and $(NH_4)_2SO_4$ solution was continuously fed to the broth. In this particular bioprocess the sugar concentration was kept low whereas the concentration of $(NH_4)_2SO_4$ and hence the conductivity steadily increased. Therefore changes in oxygen solubilities depend practically only on the increasing salt concentration. Figure 16 compares calculated and volumetrically determined oxygen solubilities in this culture medium. The temperature of 33 °C given in Ref. [12] seems to be incorrect; it should probably be about 23 °C.

It should be stressed that only the coefficients (a_0, a_1, a_2) of $(NH_4)_2SO_4$ were used for the solubility calculations as this salt was present in excess. Figure 16 shows that a striking agreement between predictions and measurements was found for this bioprocess.

It can be assumed that the method of Popović et al. [12] for estimating the O_2 solubility decrease by salts successfully applies to all microbial cultivations provided an unambiguous relation between the concentration of electrolytes and the conductivity of the culture media is available. This is the case if a certain salt is present in large excess and if the relative composition of salts present in the medium does not vary during cultivation. However, in such cases where the concentration ratios of the salts present varies and hence the set of coefficients is no longer valid the method may give only rough estimates.

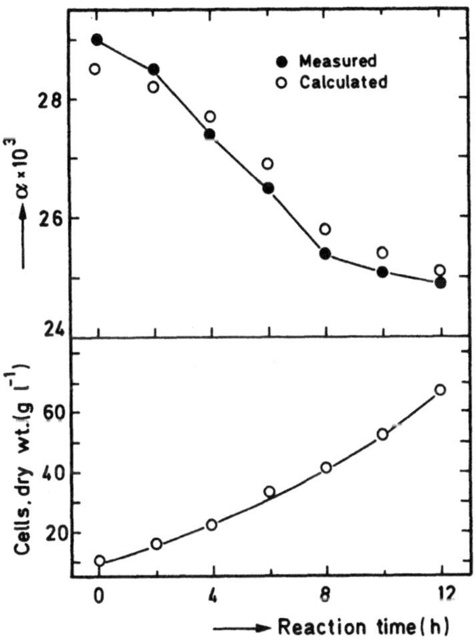

Fig. 16. O_2 solubility for growth of *Saccharomyces cerevisiae* (Popović et al. [12])

7.3 Failure of Predictive Methods

For cultivation media which consist mainly of salts and carbohydrates the oxygen solubility during actual bioreaction can be predicted rather reliably as was shown in the previous sections. This, however, appears doubtful if the main components of the media are proteins and albumens. As an example, Fig. 17 gives the oxygen solubility, the glucose concentration and the amount of NaOH added per liter of reaction mixture as a function of the reaction time of *Escherichia coli* [73]. The nutrition medium consists mainly of proteins and the exact composition is given in Table 11 (No. XIII). Drastic but reproducible solubility changes can be observed which cannot be explained by the methods outlined above. For instance, the decrease of the glucose concentration cannot be used to describe the increase in oxygen solubility during the first period of bioprocess. Even if a sharp decrease of the NaCl concentration present in the culture medium is assumed the solubility increase cannot be explained. The observed changes in solubility are probably caused by variations of the protein structure and accompanying adsorption effects. It is understood that the method of Popović et al. [12] cannot be applied either to cultivations like the one

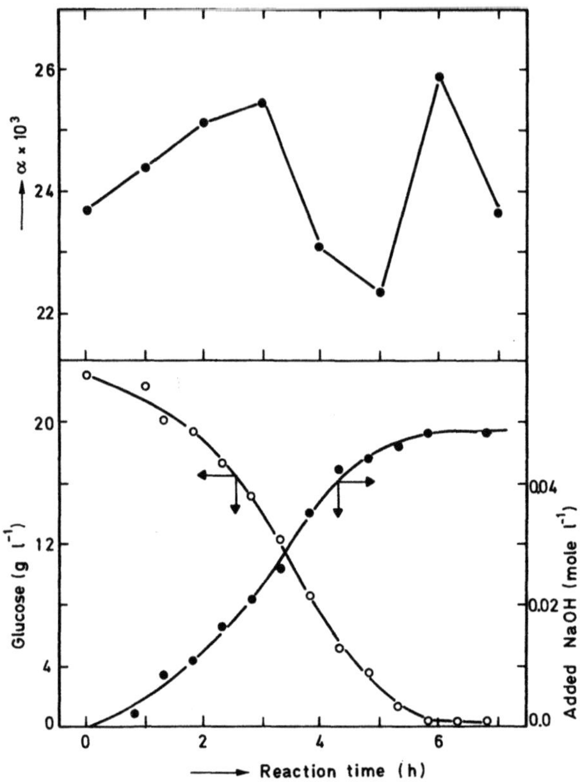

Fig. 17. O_2 solubility, glucose concentration and added amount of NaOH for cultivation of *Escherichia coli* at 28 °C (Quicker [73])

presented in Fig. 17. In such cases it is recommended to apply the experimental techniques described in Sect. 3.

8 Concluding Remarks

Since gas solubilities in liquids are usually small the effects of various dissolved salts and organic compounds on them can be accounted for by a semi-empirical log-additivity law, i.e., Eqs. (40) and (41), respectively. The analysis of solubility data and the results presented in this paper demonstrate that the direct predictive approach works successfully if the oxygen solubility is predominantly influenced by the presence of salts and carbohydrates provided their concentration changes during cultivation can be estimated. In such cases the method of Popović et al. can also be applied, particularly, if the solubility is mainly governed by one or several surplus electrolytes. The application of both methods can be recommended while taking into consideration their specific limitations.

If the media consist mainly of proteins and albumens, changes in gas solubility during cultivation are observed which cannot yet be explained reasonably. They are probably due to changes in the protein structure and accompanying adsorption phenomena. Further careful work is needed to clarify these effects. In addition, the effect of small amounts of alcohols on oxygen solubility and the confusing findings with various solids and possible adsorption effects on biomass urgently require thorough experimental investigation and analysis.

9 Acknowledgments

The authors gratefully acknowledge the financial support for the experimental work from the Ministry of Research and Technology of the Federal Republic of Germany. Thanks are due to Prof. Dr. K. Schügerl and his biotechnology group (Universität Hannover) for their cooperation by providing various culture media.

10 Nomenclature

c	mass concentration, mg l^{-1}
c_{el}	electrolyte (salt) concentration, mole l^{-1}
c_g	concentration in gas phase, mg l^{-1}
c_j	solubility of gas j, mole l^{-1}
c_{j0}	solubility of gas j in water, mole l^{-1}
c_l	concentration of solute l, mole l^{-1} or g l^{-1}
c_m	molarity, mole l^{-1}
c_n	concentration of nonelectrolyte, g l^{-1}
c_w	weight solubility, mole g^{-1}
f_j	activity coefficient of solute gas j
f_{j0}	activity coefficient of solute gas j in water
H_i	salting-out parameter of ion i, Eq. (34), l $mole^{-1}$
H_c	Henry's constant, Eq. (13), kPa l mg^{-1}
H_L	Henry's constant, Eq. (4)
H_m	Henry's constant, Eq. (2), kPa l $mole^{-1}$
H_x	Henry's constant, Eq. (3), kPa
h	empirical parameter, Eq. (27), l $mole^{-1}$
h_+, h_-, h_G	empirical parameters of van Krevelen-Hoftijzer model referring to cation, anion and gas, l $mole^{-1}$
I	ionic strength, mole l^{-1}
I_i	ionic strength of single ion, Eq. (35), mole l^{-1}
k_j	constant specific of gas j, l $mole^{-1}$
k_s	constant specific of salt, l $mole^{-1}$
K	solubility parameter for nonelectrolytes (organic compounds), Eq. (37), l g^{-1}
K_s	Sechenov constant, l $mole^{-1}$
L	Ostwald coefficient
m	parameter defined by Eq. (38), l g^{-1}
M_G	molecular weight of gas, g $mole^{-1}$
M_{st}	molecular weight of solvent, g $mole^{-1}$
p	partial pressure, kPa

P_s	vapor pressure of solvent, kPa
P_{tot}	total pressure, kPa
r	oxygen uptake rate, mg l^{-1} s^{-1}
R	rate of oxygen partial pressure decrease, kPa s^{-1}
R	gas constant, kPa cm^3 $mole^{-1}$ K^{-1}
S	Kuenen coefficient, cm^3 g^{-1}
t	temperature, °C
T	temperature, K
V_0	molar volume of gas, cm^3 $mole^{-1}$
V_G	gas volume, cm^3
V_L	liquid volume, cm^3
w	mass fraction
x	mole fraction
x'	electrical conductivity, Ω^{-1} cm^{-1}
x_i	number of ions of type i in electrolyte

Greek letters

α	Bunsen coefficient
α_0	Bunsen coefficient of water
β	absorption coefficient
ϱ_s	density of solution
ϱ_{st}	density of solvent

11 References

1. Hitchman, M. L.: Measurement of Dissolved Oxygen. John Wiley & Sons, Inc. and Orbisphere Corp., Geneva and York 1978
2. Lee, Y. H., Tsao, G. T.: Adv. Biochem. Eng. *13*, 35 (1979)
3. Battino, R., Clever, H. L.: Chem. Rev. *60*, 395 (1966)
4. Standard Methods for the Examination of Water and Waste Water, 13th ed., American Public Health Assoc., Amer. Waterworks Assoc. and Water Pollution Control Federation, Eds., p. 474, American Public Health Assoc., New York 1971
5. Markham, A. E., Kobe, K. A.: Chem. Rev. *28*, 519 (1941)
6. Clever, H. L., Battino, R.: In Solutions and Solubilities, (Dack, M. R. J. ed.), Techniques of Chemistry *8* (1), p. 379, Wiley, New York 1975
7. Phillips, D. H., Johnson, M. J.: J. Biochem. Microbiol. Technol. Eng. *3*, 277 (1961)
8. Liu, M. S., Branion, R. M. R., Duncan, D. W.: Biotech. Bioeng. *15*, 213 (1973)
9. Käppeli, O., Fiechter, A.: Biotech. Bioeng. *23*, 1897 (1981)
10. Lehmann, J. et al.: Poster paper presented at 6th Int. Fermentation Symp. London, Canada, July 20–25, 1980
11. Baburin, L. A., Shvinka, J. E., Viesturs, U. E.: Europ. J. Appl. Microbiol. Biotechnol. *13*, 15 (1981)
12. Popović, M., Niebelschütz, H., Reuß, M.: ibid. *8*, 1 (1979)
13. Quicker, G. et al.: Biotech. Bioeng. *23*, 635 (1981)
14. Ben-Naim, A., Baer, S.: Trans. Faraday Soc. *60*, 1736 (1964)
15. Tokunaga, J.: J. Chem. Eng. Jap. *8*, 7 (1975)
16. Kojima, K., Tochigi, K.: Paper presented at 7th CHISA, section D 2.11, Prague 1981
17. Long, F. A., McDevit, W. F.: Chem. Rev. *51*, 119 (1952)
18. Konnik, E. I.: Russ. Chem. Rev. *46*, 577 (1977)

19. Sechenov, M.: Ann. Chim. Phys. *25*, 226 (1892)
20. Debye, P., McAuley, J.: Phys. Z. *26*, 22 (1925)
21. Debye, P.: Z. physik. Chem. *130*, 56 (1927)
22. Bockris, J. O'M., Bowler-Reed, J., Kitchener, J. A.: Trans. Faraday Soc. *47*, 184 (1951)
23. Conway, B. E., Desnoyers, J. E., Smith, A. C.: Phil. Trans. Roy. Soc. London *A256*, 389 (1964)
24. Ruetschi, P., Amlie, R. F.: J. Phys. Chem. *70*, 718 (1966)
25. McDevit, W. F., Long, F. A.: J. Am. Chem. Soc. *74*, 1773 (1952)
26. Shoor, S. K., Gubbins, K. E.: J. Phys. Chem. *73*, 498 (1969)
27. Masterson, W. L., Lee, T. P.: ibid. *74*, 1776 (1970)
28. van Krevelen, D. W., Hoftijzer, P. J.: Chimie et Industrie; p. 168, Numéro Spéciale du XXIe Congrés International de Chimie Industrielle, Bruxelles 1948
29. Prytz, K., Holst, H.: Ann. Physik *54*, 130 (1895)
30. Usher, F. L.: J. Chem. Soc. *97*, 66 (1910)
31. Morgan, J. L. R., Pyne, H. R.: J. Phys. Chem. *34*, 1578 (1930)
32. v. Kiss, A., Lajtai, I., Thury, G.: Z. anorg. Chem. *233*, 346 (1937)
33. Curry, J., Hooselton, C. L.: J. Am. Chem. Soc. *60*, 2771 (1938)
34. Markham, A. E., Kobe, K. A.: ibid. *63*, 449 (1941)
35. Harned, H. S., Davis, R.: ibid. *65*, 2030 (1943)
36. Morrison, T. J., Billet, F.: J. Chem. Soc. p. 3819 (1952)
37. Gjaldbaek, J. C.: Acta Chem. Scand. *7*, 537 (1953)
38. Bartholme, E., Fritz, H.: Chem.-Ing.-Tech. *28*, 706 (1956)
39. Novak, J., Fried, V., Pich, J.: Coll. Czech. Chem. Commun. *26*, 2266 (1961)
40. Yeh, S.-Y., Peterson, R. E.: J. Pharm. Sci. *53*, 822 (1964)
41. Murray, C. N., Riley, J. P.: Deep Sea Res. *18*, 533 (1971)
42. Perez, J. F., Sandall, O. C.: J. Chem. Eng. Data *19*, 51 (1974)
43. Yasunishi, A., Yoshida, F.: ibid. *24*, 11 (1979)
44. Perry, R. H., Chilton, C. H.: Chemical Engineers Handbook, 5th ed., McGraw-Hill, New York 1973
45. Wilhelm, E., Battino, R., Wilcock, R. J.: Chem. Rev. *77*, 223 (1977)
46. Danckwerts, P. V.: Gas-Liquid Reactions, McGraw-Hill, New York 1970
47. Onda, K. et al.: J. Chem. Eng. Jap. *3*, 18 (1970)
48. Onda, K. et al.: ibid. *3*, 137 (1970)
49. Schumpe, A., Adler, I., Deckwer, W.-D.: Biotech. Bioeng. *20*, 145 (1978)
50. Schumpe, A., Deckwer, W.-D.: ibid. *21*, 1075 (1979)
51. Geffcken, G.: Z. phys. Chem. *49*, 257 (1904)
52. Winkler, L. W.: Z. angew. Chem. *24*, 341, 831 (1911)
53. MacArthur, C. G.: J. Phys. Chem. *20*, 495 (1916)
54. Eucken, A., Hertzberg, G.: Z. physik. Chem. *195*, 1 (1950)
55. Bruhn, G., Gerlach, J., Pawlek, F.: Z. anorg. allgem. Chem. *337*, 68 (1965)
56. Davis, R. E., Horvath, G. L., Tobias, C. W.: Electrochimica acta *12*, 287 (1967)
57. Khomutov, N. E., Konnik, E. I.: J. Phys. Chem. U.S.S.R. *48*, 359 (1974)
58. Yosunishi, A.: J. Chem. Eng. Jap. *10*, 89 (1977)
59. Findlay, A., Shen, B.: J. Chem. Soc. *101*, 1459 (1912)
60. Markham, A. E., Kobe, K. A.: J. Am. Chem. Soc. *63*, 1165 (1941)
61. Sada, E., Kito, S., Ito, Y.: Adv. Chem. Ser. *155*, 374 (1976)
62. Lang, W.: private communication
63. Lang, W., Wolf, H. U., Zander, R.: Anal. Biochem. *92*, 255 (1979)
64. Solubility Data Series, Pergamon Press, Oxford
65. Tokunaga, J.: J. Chem. Eng. Data *20*, 4 (1975)
66. Shchukarev, S. A., Tolmacheva, T. A.: J. Structural Chem. *9*, 16 (1968)
67. Lubarsch, O.: Wied. Ann. *37*, 524 (1889)
68. Schläpfer, P., Andykowski, T., Bukowiecki, A.: Schweizer Arch. Angew. Wiss. Tech. *15*, 299 (1949)
69. Findlay, A., Shen, B.: J. Chem. Soc. *101*, 1459 (1912)
70. Koch, B.: Diploma work, Univ. Hannover 1979
71. Hikita, H., Asai, S., Azuma, Y.: Can. J. Chem. Eng. *56*, 371 (1978)

72. Zander, R.: Z. Naturforsch. *31c*, 339 (1976)
73. Quicker, G.: Diploma work, Univ. Hannover 1980
74. Sada, E., Kito, S., Ito, Y.: J. Chem. Eng. Jap. *7*, 57 (1974)
75. Showalter, H. A., Ferguson, J. B.: Can. J. Res. *14B*, 120 (1936)
76. Christoff, A.: Z. physik. Chem. *53*, 321 (1905)
77. Usher, F. L.: J. Chem. Soc. *97*, 66 (1910)
78. Müller, C.: Z. physik. Chem. *81*, 483 (1912)
79. Wishnia, A.: Proc. Natl. Acad. Sci. U.S. *48*, 2200 (1962)
80. Wetlaufer, D. B., Lovrien, R.: J. Biol Chem. *239*, 596 (1964)
81. Wishnia, A., Pinder, T.: Biochem. *3*, 1377 (1964)
82. Wetlaufer, D. B. et al.: J. Am. Chem. Soc. *86*, 508 (1964)
83. Wishnia, A.: J. Phys. Chem. *67*, 2079 (1963)
84. Matheson, I. B. C., King, A. D., Jr.: J. Colloid Interface Sci. *66*, 464 (1978)

Reaction Engineering Parameters for Immobilized Biocatalysts

Klaus Buchholz
Dechema-Institut, Postfach 97 01 46
D-6000 Frankfurt am Main 97, FRG

The article concentrates on those biochemical engineering parameters of immobilized biocatalysts which are considered important with respect to their application in industrial processes. Thus swelling behaviour, mechanical stability, pressure drop and abrasion are stressed. The effectiveness is discussed in the context of external mass transfer, pore diffusion and enzyme kinetics, including systems with two substrates and two enzymes. Properties which affect the operational stability are summarized. Selected data from the recent literature are included. New concepts directed towards an optimized catalyst design are discussed, emphasizing the effectiveness and productivity.

1 Introduction

Early successful experiments on the immobilization of active proteins [1,2] did not draw much attention to this principle which was new in heterogeneous catalysis as well as in biotechnology. In the mid-sixties when much pioneering work was achieved great expectations were focussed on the technical promise of carrier-bound enzyme systems, which in turn resulted in a somewhat disappointed climate in the mid-seventies.

In fact a considerable number of processes are by now well established on an industrial scale [3]. Among these the conversion of glucose to fructose by glucose isomerase — its scale being near 2 million tons of product per year — and the hydrolysis of benzyl penicillin by penicillin acylase are the most important examples.

One area of recent research on immobilized biocatalysts is directed towards the optimization of those first generation, one-step enzyme reactions. Relevant topics are the catalyst effectiveness, operational stability and more sophisticated reaction engineering. The aim is to gain an advanced insight into the basic phenomena involved as well as technical progress.

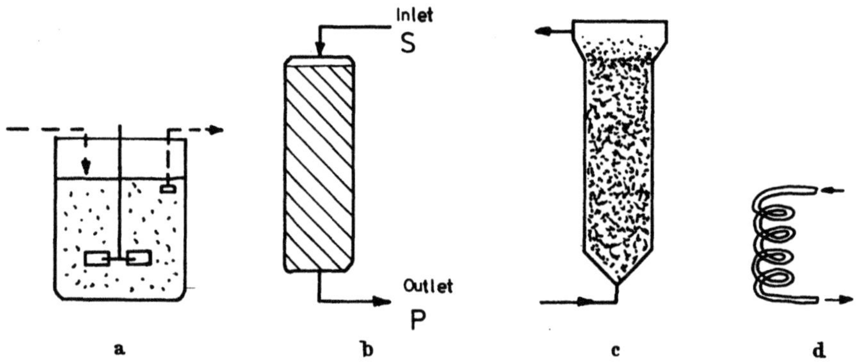

Fig. 1a–d. Most common types of reactors used for immobilized heterogeneous biocatalysts (cf. Ref. 4p. 7). **a** stirred tank, **b** fixed bed, **c** fluidized bed, **d** tubular reactor (with enzymes attached to the wall of the tube)

Table 1. Relevant parameters affecting the performance of heterogeneous immobilized biocatalysts [5]

	Physical and chemical parameters involved[1]
Swelling behaviour of the carrier	pH, I
Mean wet particle diameter (d_p) (distribution, shape)	pH, I
Particle compression behaviour,	Δp
flow resistance in fixed beds,	Δp, h, u, ν, d_p
abrasion in stirred vessels	d_i, n
fluidization velocity	d_p, u, $\Delta\varrho$, ν
Maximum activity, or initial reaction rates (V_{max}, v)	S, P, t, T, pH, I and buffer conc., d_p, u or d_i, n
Effectiveness η as a function of	
external mass transfer,	d_p, ν, u, or d_i, n
pore diffusion,	d_p, V_{max}, K_M, T, D_e, S, P
partition effects,	
degree of conversion	
Operational stability, depending on abrasion, enzyme inactivation, fouling, irreversible adsorption, occlusion etc.	t, S, P, T, pH, conc. of other compounds

[1] Symbols are given at the end of the article

A second area of research and development deals with more complex systems, e.g. those requiring coenzymes and/or sequential enzymatic reactions where a break-through on the industrial scale has not yet been accomplished.

Immobilized biocatalysts comprise both enzymes and cells immobilized by attachment to soluble polymers, insoluble carriers or by entrapment in membrane systems. This article concentrates on the first area mentioned above, dealing with enzymes or cells bound to insoluble porous carriers which have been used in stirred vessels or fixed bed reactors (Fig. 1), as well as on reaction engineering aspects associated with heterogeneous biocatalysts. These aspects comprise the physical and mechanical parameters as well as transport phenomena and kinetics which are intimately linked to biochemical aspects. The most important parameters are summarized in Table 1. These parameters have been discussed in more detail together with aspects of catalyst synthesis and test methods by a research group [5], and they were recently emphasized and reduced in number by a European working party [6].

A better insight into the phenomena mentioned provides a basis for optimization in catalyst synthesis and reactor design, and examples of such efforts will be given. The methods of immobilization have been summarized many times, and excellent reviews are available (e.g. [7-10]).

2 Mechanical and Physical Parameters of Carrier-bound Biocatalysts

Most heterogeneous biocatalysts applied in technical processes are particulate, where enzymes or cells are immobilized inside porous granular or spherical carriers. It is obvious that the physical characteristics of the matrices will be of major importance for the performance of such systems under technical conditions. Thus the application range in stirred tanks, fluidized and fixed beds depends on the mean diameter of catalyst particles and its distribution, their density, swelling behaviour, their mechanical strength or compression behaviour (for test methods see 5).

2.1 Particle Diameter

The particle diameter d_p and its distribution is of major importance for suspending the catalyst in stirred tanks or fluidized beds, for the pressure drop in fixed beds, and for the effectiveness under given experimental conditions. It is obvious that they must be determined under application conditions (pH, I). The most simple method is the observation under the microscope and/or sieving (c.f. Fig. 2). Other methods are light scattering techniques or the light blockage principle [11]. Sieving in the wet state is generally recommended for biocatalysts with heterogeneous dimensions, since both their mechanical performance in fixed bed reactors and their biochemical activity and efficiency strongly depend on the particle size distribution. The size of industrial biocatalysts ranges from 0.2 to 4 mm, where immobilized enzyme systems are normally found in the lower and immobilized cells in the upper range, corresponding to high or low activities per carrier volume, respectively [3].

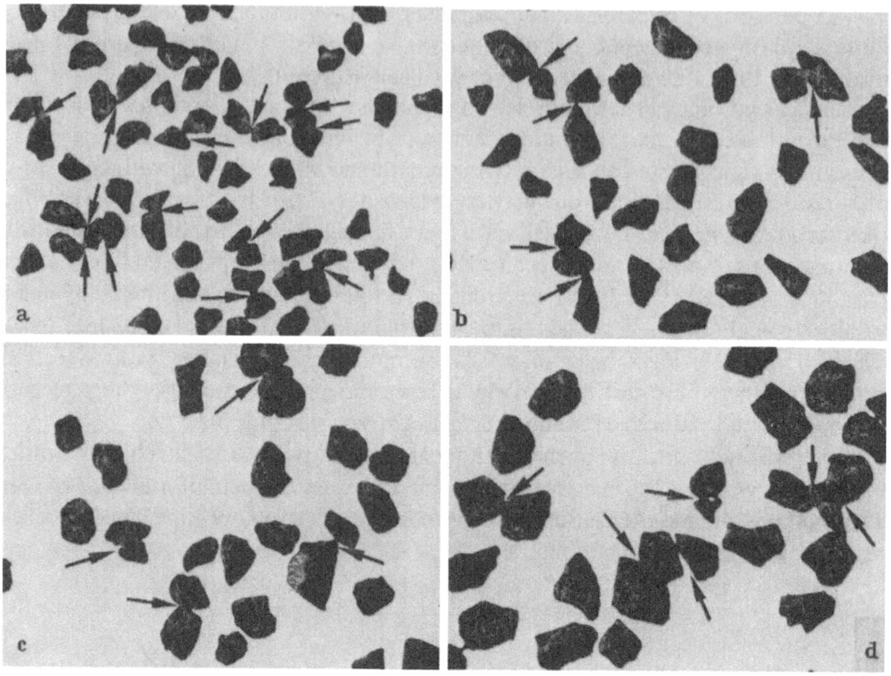

Fig. 2a—d. Photomicrographs of particle fractions obtained by wet sieving [11]. **a** < 100 μm; **b** 100 to 125 μm; **c** 125–160 μm; **d** 160–200 μm

2.2 Swelling Behaviour

The swelling behaviour of a carrier determines the density of active catalytic sites inside the actual reaction volume of the catalyst. It can also serve as a rough indicator of the compressibility of the particles which is high in most cases where strong swelling ($10\ ml\ g^{-1}$) is observed. It should be determined from the wet weight or settled volume of the carrier in a solution corresponding to application conditions [5a)]. Furthermore, the density of the particles plays a role in fluidized bed application and for the external mass transfer.

2.3 Pressure Drop and Particle Compression Behaviour

The pressure drop is of major importance for the application of biocatalysts in fixed bed reactors. These may be several meters in height with pressure drop above one bar [3,12)]. A general correlation for the pressure drop Δp as a function of the bed height h, flow rate u, voidage ε and particle diameter d_p is given by Eq. (1) for laminar flow conditions (when $Re = ud_p/\nu \leq 10$) [13)]:

$$\frac{\Delta p}{h} = 150 \frac{(1-\varepsilon)^2 \eta u}{\varepsilon^3 d_p^2} \tag{1}$$

For turbulent flow conditions this equation must be extended [13]:

$$\frac{\Delta p}{h} = 150 \frac{(1-\varepsilon)^2 \eta u}{\varepsilon^3 d_p^2} + 1{,}75 \frac{(1-\varepsilon)\,\varrho u^2}{\varepsilon^3 d_p} \tag{2}$$

It is obvious from Eq. (1) that both the mean particle diameter (d_p) and the voidage (ε), — which in turn depends on the distribution of d_p — determine the pressure drop of fixed beds. The voidage (ε) — defined as the interstitial space not occupied by the catalyst particles devided by the total volume of the bed — commonly varies from 0.3 to 0.5. For very heterogeneous or compressible carriers it may fall below 0.3. ε can be determined with high molecular weight markers, e.g. dextran derivatives [5b].

The experimental determination of the pressure drop is strongly recommended for fixed bed application. A rather simple method applying varying flow rates is shown in Fig. 3. Corrections are necessary if the assembly contains screens and capillaries. The results from such measurements may also serve for estimating the pressure drop at different bed heights when laminar flow conditions are maintained, where the dependence of Δp on u and h is equivalent [Eq. (1)].

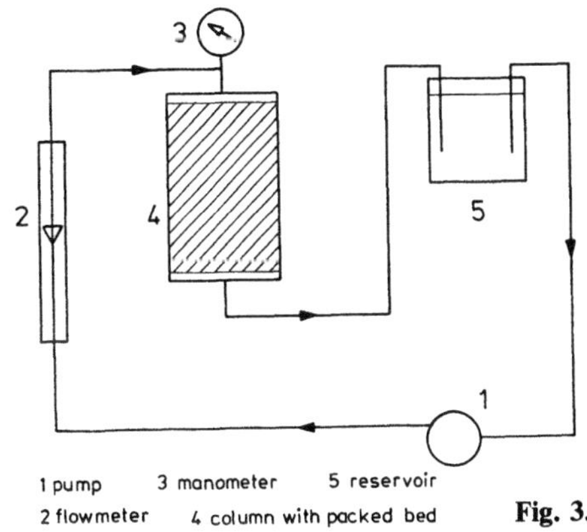

Fig. 3. Determination of the pressure drop in a fixed bed

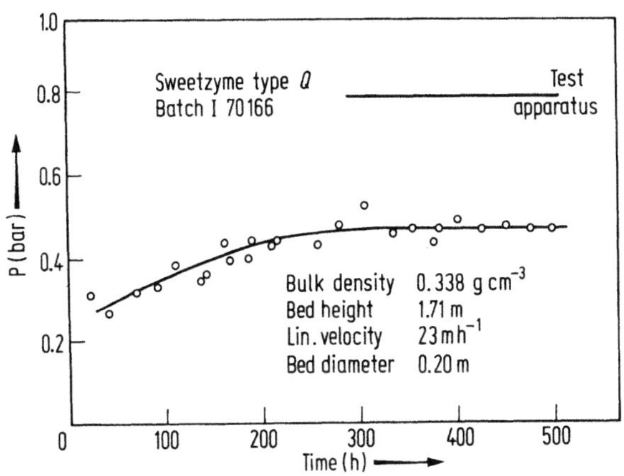

Fig. 4. Time-dependence of the pressure drop in a fixed bed reactor [14]

Extrapolation from laboratory results to the technical or industrial scale is difficult. For this purpose a more sophisticated apparatus has been developed [14]. The dependence of the pressure drop on operation time is shown in Fig. 4. An empirical equation has been presented for correlating the pressure drop with the flow rate and bed height and which includes a time correction [14].

It is notably with carriers which undergo compression or deformation that only experimental investigations provide relevant information on the hydrodynamic behaviour of the catalyst. Equation (1) is not valid in such cases where the pressure

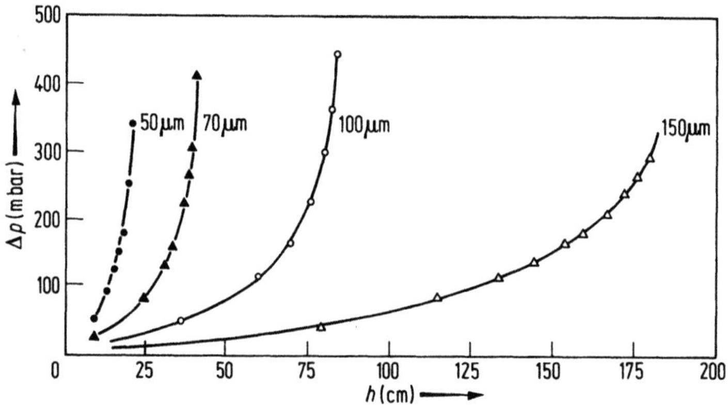

Fig. 5. Pressure drop as a function of bed height in a fixed bed reactor with compressible or deformable particles of different diameters; model calculation [15]

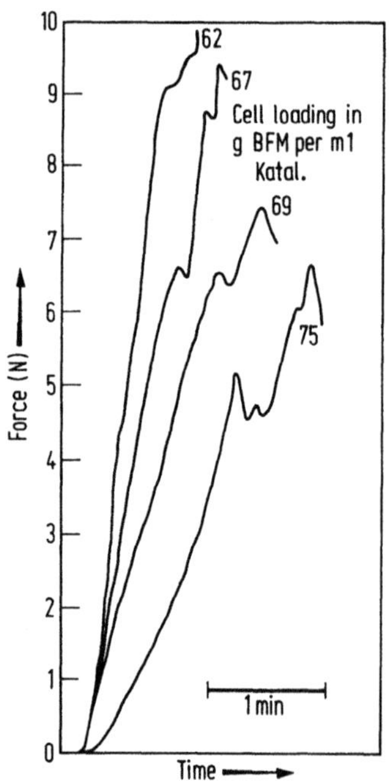

Fig. 6. Particle compression experiment. Epoxy-beads are compressed by two plates moving at a given rate. The particles are broken at the inflection point. The cell loading (cell wet weight per ml of catalyst particle) is indicated [16]

drop exhibits a nonlinear correlation with flow rate (or bed height) at laminar flow conditions. This is shown in Fig. 5 [15] giving calculated results similar to experimental findings. They are based on a model which takes a deformation modulus for catalyst particles into consideration. The strong influence of the particle size on pressure drop will be pointed out.

The phenomena described depend on the mechanical characteristics of particles which can be investigated by experiments where a single particle is submitted to compression between two plates provided with a driving motor and a force measuring cell [16]. The results provide information on the elastic behaviour of carriers and on the critical compression force at which a particle will break (Fig. 6). This figure also shows that the mechanical strength decreases with increasing cell loading for an immobilized cell system. From such measurements it has been shown

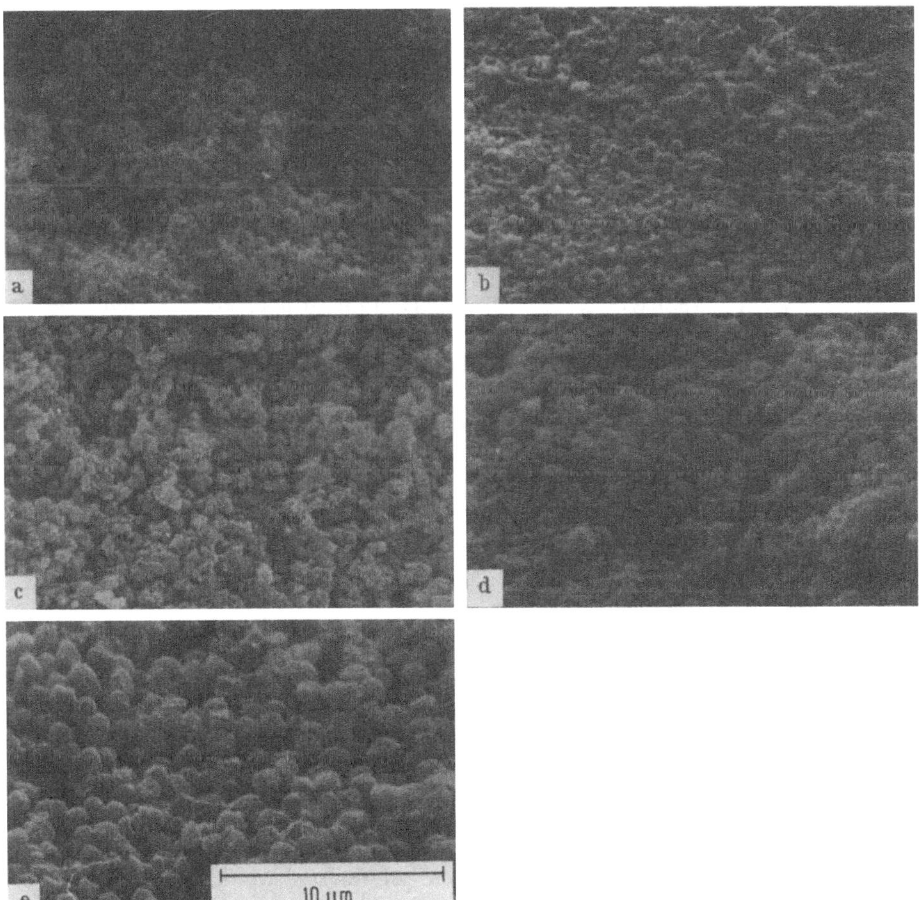

Fig. 7a—e. Electron micrographs of macroreticular carriers for enzyme immobilization [18b]; pore structures obtained with different degrees of crosslinking (carriers; 4-isothiocyanatostyrene/acrylic acid/ 1,4 divinylbenzene 1 :10 : x) (scanning electron microscopic-micrographs: Dr. Hoder, FU Berlin). **a** x = 2.20; **b** x = 1.65; **c** x = 1.10; **d** x = 0.66; **e** x = 0.44

that epoxy-, dried Ca-alginate- and chitosan carriers exhibit good mechanical properties [16,17]. Macroreticular polymers which have been developed for enzyme immobilization and high binding capacity also provide excellent mechanical stability which is due to their structural features [18,19]. Figure 7 shows electron micrographs of macroreticular particles composed of rigid microspheres.

2.4 Abrasion

Tests should be performed in baffled stirred vessels at a Reynolds number ($Re = \pi n d_i^2 v^{-1}$) comparable to normal operation conditions (recommendation: $Re > 50000$, e.g. $n = 500\ min^{-1}$, $d_i = 5$ cm). Only minor effects have been found with carriers for enzyme immobilization [20]. For immobilized cell systems, very pronounced effects were found to be associated with increasing cell loading, notably in the range >60% with particle size >1 mm and with high volume fractions of carrier in the reaction vessel (30%) [21].

3 Kinetics and Effectiveness

This topic has been reviewed repeatedly (c.f. [22–24]). The effectiveness is one of the most important parameters for the characterization of a catalytic system and its performance with respect to application and economics. This chapter will summarize those aspects which are considered most important and include subjects which were dealt with in the recent literature, e.g. operational effectiveness and systems with more than one substrate or enzyme.

3.1 Basic Considerations

Immobilization Efficiency

This parameter is of major importance for the economics of a biocatalyst. Both the amount of bound protein or cell mass and active immobilized enzyme should be analyzed. The first can simply be calculated from the difference of protein offered in the immobilization procedure and that recovered in the solution and washing solutions. A more accurate method is based on the hydrolysis of the catalyst and amino acid analysis [25,26]. Active immobilized enzyme is determined via the kinetics. With rather stable enzymes, e.g. hydrolytic ones, common results are in the range of 20—50 mg ml^{-1} immobilized protein, 50—90% binding yield and 25—80% yield of active immobilized enzymes based on enzyme offered for binding (c.f. [5]). For industrial biocatalysts yields even higher than 90% were reported [27]. Cell immobilization values for the loading are in the range from 10% to 75% (cell wet weight per g of catalyst) [16,28,29]. Relative activities, based on the amount of immobilized cells, are typically in the range of 30—80%, where inactivation and mass transfer may play a role [28,30].

Phenomena Involved in Macrokinetics

Rate controlling steps and important phenomena are (c.f. Fig. 8):

External mass transfer of substrates and/or products between the well-mixed bulk fluid and the surface of carriers;

Partition effects at the fluid-carrier interface, especially important with ionic substrates and/or products and ion exchange carriers;

Pore diffusion of substrates and/or products;

Catalytic reaction(s) with intrinsic kinetics.

The effectiveness of a given catalyst under certain boundary conditions depends on the limitation or influence of these individual steps on the overall reaction. It is difficult however to determine their respective weight since only overall reaction rates (macrokinetics) can be determined for a given catalyst.

Accessible experimental information includes in general:

Initial reaction rates, as a function of concentrations, pH, temperature;

The maximum reaction rate (in those cases where simple Michaelis kinetics apply for the enzyme and substrate solubility is sufficiently high);

Conversion versus residence time curves.

The determination of the catalyst effectiveness may need much more experimental investigation as well as mathematical treatment, depending on the complexity of the system.

The *effectiveness* of a catalyst relates the overall reaction rate for a given immobilized enzyme or cell system to that of the same amount of native biocatalyst under otherwise identical conditions (concentrations etc., see Table 1):

$$\eta = v_{(imm)}/v_{(native)} = f(Sh, \varphi, S_i) \quad (3)$$

It depends most significantly on the Sherwood number, Sh, which characterizes the external mass transfer, and the Thiele modulus φ and substrate concentrations S_i, which characterize the coupled pore diffusion and reaction kinetics. It can be determined in a simple way only if the amount of active immobilized enzyme inside the carrier is known (e.g. by active site titration). Notably values for η at different substrate concentrations are of interest when referring to application conditions. That substrate concentration which corresponds to half the maximum reaction rate has

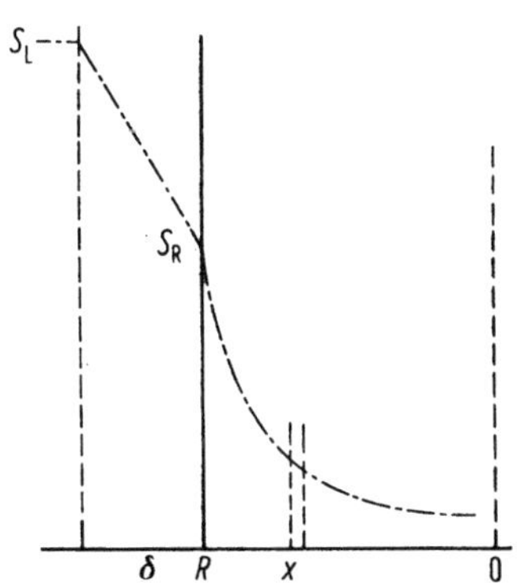

Fig. 8. Simplified scheme for coupling of mass transfer and reaction: substrate concentration profiles in the bulk phase solution (S_L), external stagnant fluid film with thickness δ, and inside the porous catalyst with radius R

found special consideration — $K_{M(app)}$ — including theoretical approaches [31,32]. The evaluation of $K_{M(app)}$ as a result of linearized plots based on Michaelis kinetics (Lineweaver-Burk or Eadie-Hofstee) must be considered with care, since any interpretation of heterogeneous systems by homogeneous kinetics may lead to erroneous results. Even papers published recently use such misleading approaches. Much more useful are plots of conversion versus time, when the experimental details are given completely.

3.2 Temperature Dependence

The temperature dependence of chemical reactions like enzymatic ones is in general pronounced (40—60 kJ per mol) in contrast to transport processes. Thus it may serve as an indicator for mass transfer limitation when the experimental temperature dependence is significantly lower than that of the homogeneous enzymatic reaction [33-36]. An example is shown in Fig. 9 where the activation energy decreases with increasing temperature, indicating increasing mass transfer limitation. It is also obvious from Fig. 9 that this effect is more pronounced with fixed beds of the catalyst where the influence of the external mass transfer will be more significant, as compared to stirred vessels.

The phenomenon depends on the substrate concentration if Michaelis kinetics apply, as can be seen from the results of Table 2. These effects may be explained easily from model calculations [33]. At high substrate concentrations ($S \gg K_M$) all active centres will be supplied with sufficient substrate for operating at V_{max}, even when external mass transfer is poor. So the temperature dependence will correspond to that of V_{max}. At lower substrate concentrations not all of the active centers of the enzymes are fed with sufficient substrate so that the mass transfer (with low activation energy) will be rate limiting, especially when the external mass transfer is

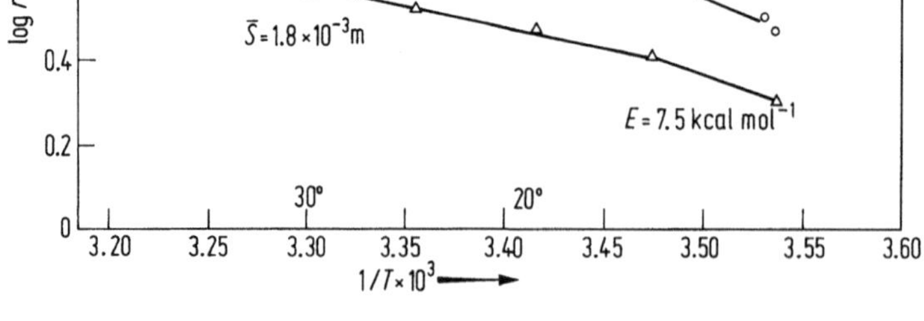

Fig. 9. Temperature-dependence of the hydrolysis of benzoyl arginine ethyl ester by immobilized trypsin (ordinate: logarithm of the reaction rate), above: stirred vessel; below: fixed bed [33]

poor [decreasing substrate concentration over the external liquid film (boundary layer)].

Table 2. Apparent activation energies E′ at different substrate concentrations (hydrolysis of benzoyl-arginine ethyl ester by immobilized trypsin, buffered solution at pH-optimum) [33]

S_L (mol l^{-1})	E′ (kJ per mol)
3×10^{-2}	40
2×10^{-2}	38
5×10^{-3}	33
2×10^{-3}	29
2×10^{-4}	27
10^{-4}	19
(experimental error ± 4 kJ per mol)	

3.3 External Mass Transfer

The phenomena involved are conveniently discussed with the aid of the simplified film model (Fig. 8) assuming a stagnant layer of fluid which adheres to the outer catalyst surface. Substrates and products must penetrate this film by diffusion according to their concentration profile. The rate of this mass transfer step is described by

$$k_s(S_L - S_R) = (D/\delta)(S_L - S_R) \quad (4)$$

In the steady state it is equal to the overall reaction rate:

$$v_c = k_s F(S_L - S_R)/V_c \quad (\text{mol}\, l^{-1}\, s^{-1}) \quad (5)$$

k_s can be determined experimentally (whereas δ does not have a real physical significance for most experimental conditions). It is strongly correlated with the hydrodynamics of the reactor system, e.g.:
the flow rate in fixed and fluidized beds;
the stirrer diameter and speed in stirred vessels;
the viscosity of the fluid.

Correlations for mass transfer and hydrodynamics are available from the chemical engineering literature. However, the fundamentals of stirred vessel hydrodynamics are not fully understood and very few experimental results have been published for immobilized biocatalysts. Since those systems in general fall outside the parameter range of common non-biological systems (e.g. with respect to d_p, $\Delta\varrho$), values estimated from the literature by extrapolation must be applied with care.

General Correlations

The correlations for mass transfer and experimental conditions are commonly expressed in terms of dimensionless groups:

$$Sh = c\, Re^a\, Sc^b \quad (6)$$

where

$$Sh = \frac{k_s d_p}{D} \qquad Sc = D/\nu$$

$$Re = \frac{u d_p}{\nu} \qquad \text{for fixed beds}$$

$$Re_i = \frac{n d_i^2}{\nu} \qquad \text{for stirred vessels}$$

For fixed beds and laminar flow, values are reported as:

$$c = 4{,}6\,, \qquad a = b = 1/3$$ [37];

for non-laminar flow ($20 < Re < 120$):

$$c = 2{,}4\,, \qquad a \simeq 0{,}5\,, \qquad b = 1/3$$ [37].

From these correlations, Sh and k_s can be calculated. This is not true however for compressible particles, which exhibit poor external mass transfer in fixed beds.

Several investigations with immobilized enzymes in fixed beds have been published [38, 39], where also deviations from ideal behaviour (c.f. correlations given before) were observed due to channelling [40]. Axial dispersion is of minor importance

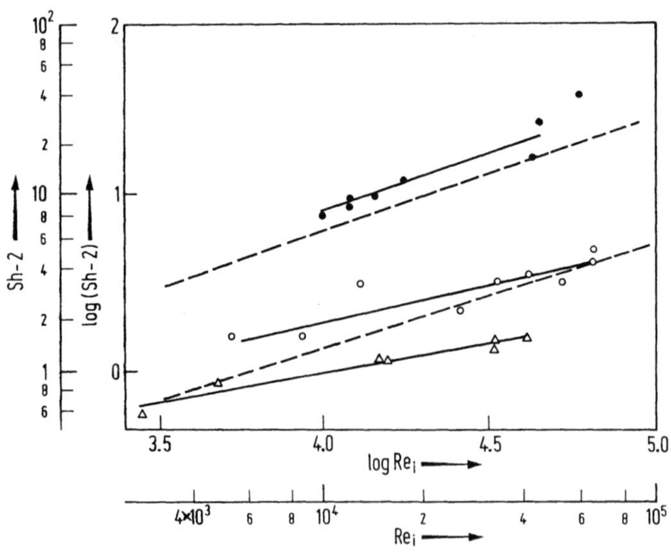

Fig. 10. Dependence of (Sh-2) on the Re number of stirred vessels, (•) porous glass, $d_p \simeq 0.1$ mm, (○) and (△) Sepharose with $d_p \simeq 0.1$ and 0.07 mm respectively; dashed lines represent correlations from [45] above and [46] below, respectively

in fixed bed reactors [23]. The dispersion coefficient in the case of immobilized invertase has been determined [41].

In fluidized bed operation the axial dispersion obviously plays a role. General correlations are given in [23]. Examples for the performance of immobilized enzyme systems have been described [42–44]. The advantages of such reactors are a low pressure drop and the absence of clogging problems, as compared to fixed beds, and less back mixing, as compared to continuous stirred tank reactors. This is especially important when product inhibition is significant [42]. The operational range depends on the particle diameter and the density difference ($\Delta\varrho = \varrho$ (particle) $- \varrho$ (fluid)). Lactase immobilized on inorganic carriers with particle diameters in the range of 0,1–0,4 mm were successfully used with flow rates from 0.5 to 4 cm s^{-1} and conversions which compared favorably with those of fixed bed reactors [42].

For stirred vessels, literature results are less reliable especially due to the properties of biocatalysts. Approximate values for Sh and k_s may be obtained from Fig. 10 which summarizes different correlations from the chemical engineering literature and experimental results with different biocatalyst systems. It should be mentioned that carriers with a small particle diameter (<0.1 mm) and density similar to that of water exhibit poor mass transfer coefficients and Sh numbers (order of magnitude 3–6) in laboratory reactors (Re $\leq$ 60,000). This may lead to severe limitation of the overall reaction rate by external mass transfer for specific boundary conditions (especially with very active catalysts and when S is in the order of magnitude of K_M or lower). The effect of partition at the external solid/fluid interface has been investigated earlier and summarized in Ref. [22]. It can be of importance for ion exchange carriers and charged substrates or products.

3.4 Intra-particle Diffusion and Reaction

3.4.1 Basic Considerations

Inside a porous catalyst the substrate has to diffuse into the carrier in order to be converted by the immobilized enzyme. At steady state a profile of substrate throughout the carrier is established (Fig. 8). Thus in most cases the effectiveness of immobilized active enzymes will be lower compared to a homogeneous system where all enzymes are supplied with substrate at its concentration in free solution.

A corresponding product profile will be established which may be affected by changes in the reaction rate and effectiveness when product inhibition plays a role. Most important is a special case in hydrolytic reactions. Protons released lead to the establishment of a pH-profile inside the carrier affecting the enzymatic reaction. This problem has been investigated experimentally [47–49] and treated theoretically by Engasser and Horvath [50].

Buffers facilitate proton transfer and thus may have a pronounced influence on experimental rates of hydrolytic reactions and on profiles of activity as a function of the pH in solution. Figure 11 shows the influence of buffer concentration on reaction rates for carriers with different activities. It is obvious that the buffer concentration should be given with results of hydrolysis reaction rates.

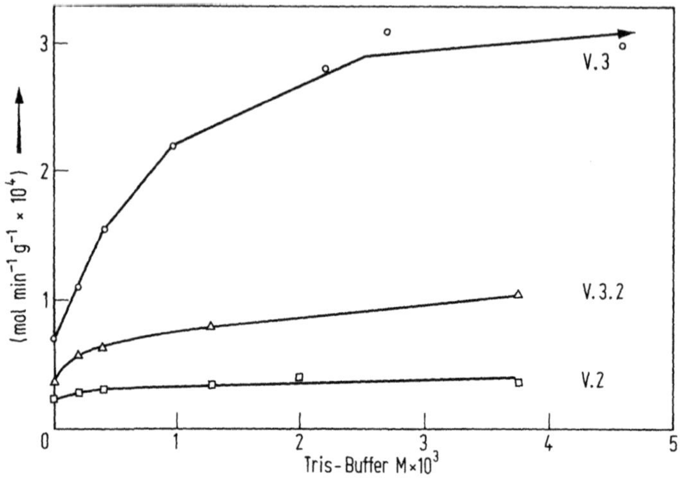

Fig. 11. Reaction rates (v_{br}) as a function of Tris-buffer concentration for porous glass with different amounts of immobilized trypsin ($d_p \simeq 100\ \mu m$, hydrolysis of benzoylarginine ethyl ester at substrate saturation, pH 8.7, 25 °C, stirred vessel, Re ≃ 13.000)

For the quantitative treatment of substrate diffusion limitation and the calculation of the effectiveness η a well established procedure is as follows:

Selection of the appropriate function $\eta = f(Sh, \varphi, S_i)$, where not all parameters must be important;

Evaluation of Sh;

Calculation or estimation of φ (according to the appropriate kinetics);

Determination of η from published tabulated or graphical correlations, or calculation by solving the differential Eq. (11), applying the mass balance (10) as boundary condition.

For convenience the mass transfer and kinetic parameters involved are summarized in appropriate dimensionless groups. Common correlations are presented in Table 3 [22–24].

The most convenient and reliable access to the effectiveness is by *experimental determination* if the amount of immobilized active enzyme can be determined (which can be difficult or impossible). It is based on the comparison of the overall reaction rate of immobilized biocatalysts, $v_{(imm)}$, to that obtained with the same amount of native enzyme or cells, $v_{(nat)}$, under otherwise identical conditions (Eq. 3). Application conditions should be selected for such measurements.

If the experimental determination is not possible, η may be calculated or estimated from the physical parameters and the kinetics of the system (Table 3). The most important *parameters* involved in intraparticle diffusion and reaction — particle diameter ($d_p = 2\ R$), kinetic constants, and effective diffusion coefficient D_e — are summarized by the dimensionless Thiele modulus φ (Eqs. 8 or 9) (attention must be paid to the definition of φ, which in some papers is different). If φ can be calculated and Sh is known η can be obtained from tabulated or graphical correlations published in the biochemical engineering literature [22–24]. For first order reactions, see Ref. [51].

Table 3. Basic correlations for the coupling of mass transfer and reaction and the calculation of effectiveness (dimensionless groups, mass balances for transport and reaction)

Effectiveness, experimental (v based on equal amounts of active catalyst in the heterogeneous and native system)

$$\eta = v_{imm}/v_{nat} \tag{3}$$

Effectiveness, theoretical (function from physical transport and kinetic parameters, summarized in dimensionless groups):

$$\eta = f(Sh, \varphi, S) \tag{7}$$

where

$$Sh = \frac{k_s d_p}{D}$$

$$\varphi = R(k_1/D_e)^{1/2} \quad \text{(first order kinetics)} \tag{8}$$

$$\varphi = R(V_{max}/K_M D_e)^{1/2} \quad \text{(Michaelis kinetics)} \tag{9}$$

Mass balance for reactions, transport through boundary liquid film and external catalyst surface (v_c based on unit volume of catalyst V_c) (boundary condition for Eq. 11):

$$v_c = k_s(F/V_c)(S_L - S_R) = D_e(F/V_c)(dS/dr)_R \tag{10}$$

Mass balance inside a volume element of the carrier:

$$D_e\left(\frac{d^2S}{dr^2} + \frac{1}{r}\frac{dS}{dr}\right) = v_r \tag{11}$$

First order reaction kinetics:

$$v_{r_1} = k_1 S_r \quad \text{(relating to volume element r)} \tag{12}$$

$$v_c = k_1' S_R = \eta_1 k_1 S_R \quad \text{(over all reaction rate, referring to unit volume of catalyst)} \tag{12a}$$

$$\eta_1 = \frac{3}{\varphi}\left(\frac{1}{\tan h\varphi} - \frac{1}{\varphi}\right), \quad \text{or approximately}$$

$$\eta_1 = 1/\varphi \quad \text{for} \quad \varphi > 5 \tag{13}$$

Michaelis kinetics[a]:

$$v = \frac{V_{max} S_r}{K_M + S_r} \tag{14}$$

$$v_c = \eta\left(\frac{V_{max} S}{K_M + S}\right), \quad \text{where} \tag{14a}$$

η is obtained from tabulated or graphical correlations, see Eq. (7) and Figs. 12, 13; or from numerical solution of Eqs. (10) and (11)

[a] If reaction rates are based on unit catalyst mass (mol s^{-1} g^{-1}) they must be converted by the catalyst density (g l^{-1}) into volume based rates

3.4.2 Simple Kinetics

First order kinetics apply for immobilized enzyme or cell systems when:
first order is a good approximation for the native system (e.g. catalase), or for Michaelis kinetics at low substrate concentration ($S_r < K_M$), which can be tested experimentally;
two substrate systems, if one substrate is rate-limiting [the other being available in excess, e.g. oxidation reactions when O_2 is rate-limiting [52], c.f. following paragraph].

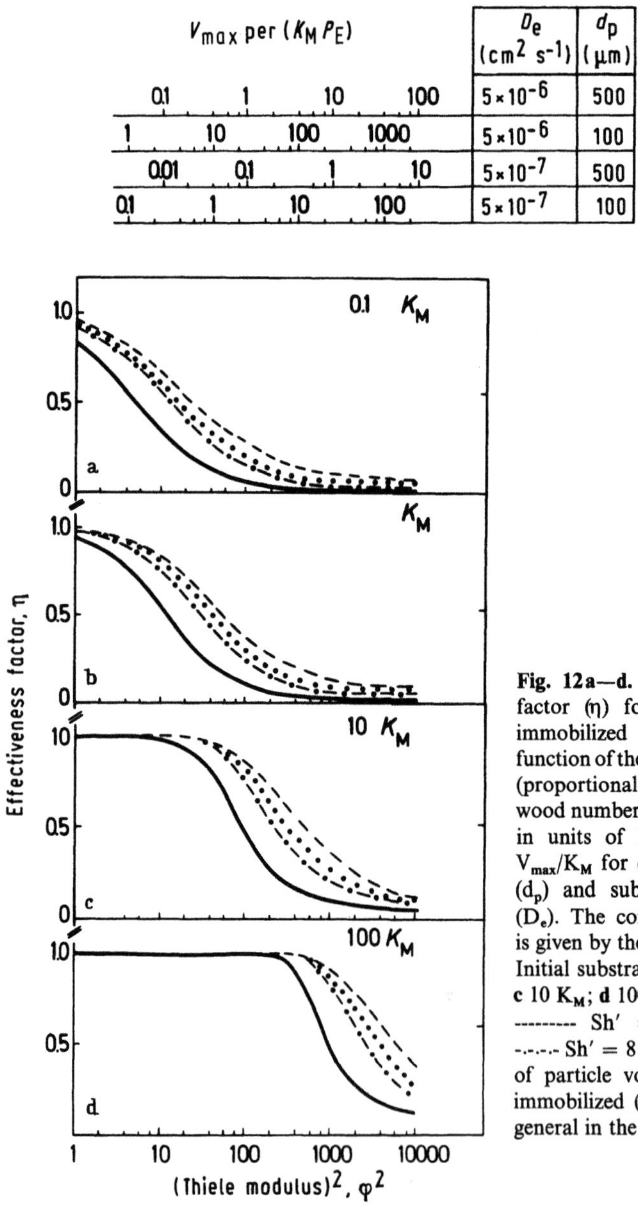

Fig. 12a—d. The stationary effectiveness factor (η) for a one substrate enzyme immobilized in a spherical particle as function of the square of the Thiele modulus (proportional to the enzyme content), Sherwood number, and initial substrate content in units of K_M. The upper scale gives V_{max}/K_M for different particle dimensions (d_p) and substrate diffusion coefficients (D_e). The corresponding Thiele modulus is given by the lower scale.
Initial substrate content: **a** 0.1 K_M; **b** K_M; **c** 10 K_M; **d** 100 K_M;
--------- Sh' = 100; ········ Sh' = 16; -.-.-.- Sh' = 8; ——— Sh' = 2. P_E fraction of particle volume where the enzyme is immobilized (fraction of pore volume, in general in the range of 0.3–0.5 [66])

The effectiveness can be calculated from Eq. (13), or estimated from Fig. 12, where $S = 0.1 K_M k_1$ can be obtained experimentally by two methods: either by measurement of the specific reaction rate (v_c) with catalyst particles of decreasing size; or by preparation of catalysts with decreasing content of enzymes or cells, in both cases down to those conditions where $k_1' = k_1 = \text{const}$ ($\eta = 1$), i.e. independent from d_p or immobilized activity.

Examples are given in Ref. [28] with cell contents in the range of 3–30% and particle diameters in the range of 0.3–0.6 mm, η varying from 0.3 to 1.

The external mass transfer is neglected in Eq. (13), which makes little difference for Sh 20. However, $k_1' = \eta_1 k_1$ can be separated from external mass transfer influence and determined by linearized plots:

$$S_L/v_c = 1/k_1' + \text{const}/n^{3/4} \tag{15}$$

where n is the stirrer speed, or

$$S_L/v_c = 1/k_1' + \text{const}'/u^{1/3} \tag{16}$$

where u is the flow rate.

Michaelis kinetics may be applied to a considerable number of systems. For the calculation of φ and η, the kinetic parameters V_{max} ($= E_0 k_c$ mol l^{-1} s^{-1}, where E_0 is the density of active enzyme in the carrier volume and k_c is the kinetic turnover rate) and K_M must be known. In general, V_{max} (not the individual k_c and E_0) is determined experimentally with substrate saturation, and the K_M value of the native enzyme is used.

In general the molecular or intrinsic kinetic parameters are unknown [22]. Their determination is inhibited by partition, adsorption and transport phenomena, which have not been excluded unambiguously in most kinetic investigations. Several investigators found that kinetic constants were not modified by immobilization [53–55], whereas unspecific effects on the protein structure could be revealed by flurescence measurements [55]. In other cases [56–59] it has been observed that the yield of immobilized enzyme activity decreased with increasing enzyme loading. However, it was not excluded that this effect could be due to diffusion limitation increasing with immobilized overall activity.

Whereas no unambiguous evidence for shifts in k_c due to immobilization exists, it has been shown that association constants for substrates and inhibitors (K_M, K_J) were considerably altered (by factors up to 5) due to the covalent attachment of an enzyme to a rigid carrier [60]. The estimation of apparent kinetic constants from coventional Lineweaver-Burk or Eadie-Hofstee-plots are not considered useful if these plots are nonlinear, since considerable errors may be associated with such approaches. In consequence it is recommended to present the original results in the form of, for example, conversion as a function of residence time for different substrate concentrations [6].

For the calculation of η on the basis of the Michaelis kinetics, approximate solutions of the differential Eq. (11) have been derived [32,62], making simplifying assumptions. Further theoretical treatments are published in Refs. [22,23,63,64].

The most convenient method for estimating η is from graphical correlations which include Sh, φ, and S_L. An example for such calculations is given in [65]. Figure 12 presents such plots for a wide range of catalyst dimensions and activities, Sh numbers and substrate concentrations. These effectiveness factors refer to constant substrate concentrations and thus apply for continuous stirred tank reactors.

For batch or continuous plug flow (tubular or fixed bed) reactors, the substrate concentration is not constant throughout the reactor and the (stationary) effectiveness varies with conversion along the reactor. For these cases, which are most important in industrial application, an *operational* effectiveness factor has been derived by Kasche [65]. It compares the times required for a given degree of substrate conversion by the same amount (based on activity) of free and immobilized biocatalyst. Corresponding graphical correlations for 90% conversion are presented in Fig. 13.

Ranges of parameters which can be found with several common experimental conditions are: For Sherwood numbers 5—20, the lower range is mostly associated with small carrier particles ($\leqq 100\ \mu m$) and low density differences, $\Delta\varrho$, between particle and fluid, the upper range with bigger particles and higher $\Delta\varrho$ (e.g. for anorganic or macroporous carriers), and for well stirred laboratory vessels (Re in the

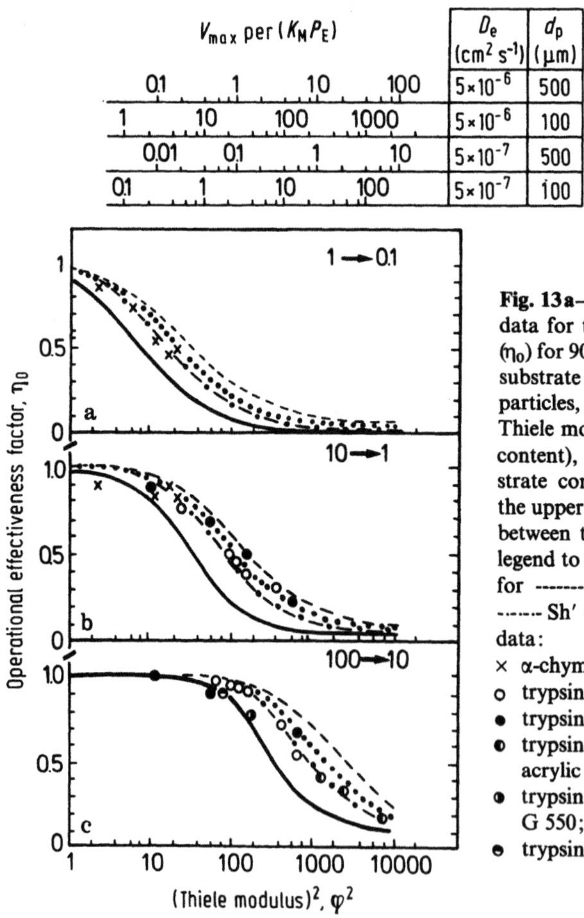

Fig. 13a—c. Calculated and experimental data for the operational effectiveness factor (η_0) for 90% substrate conversion, with a one substrate enzyme immobilized in spherical particles, as a function of the square of the Thiele modulus (proportional to the enzyme content), Sherwood number and initial substrate concentration, given in K_M-units, in the upper right hand corner. For the relation between the upper and the lower scale, see legend to Fig. 12 [66]. Curves: calculated data for -------- Sh' = 100; ········ Sh' = 16; ------ Sh' = 8; ——— Sh' = 2. Experimental data:

× α-chymotrypsin bound to Sepharose 4B;
○ trypsin bound to Sepharose 4B;
● trypsin bound to Sepharose CL 2B;
◐ trypsin bound to isothiocyanatostyrol acrylic acid matrix;
◑ trypsin bound to porous glass Servachrom G 550;
◒ trypsin bound to Oxiran 5120 B.

range of 15,000 to 60,000) (c.f. Fig. 10). For Sh = 20, the influence of external mass transfer can in general be neglected.

The effective diffusion coefficient D_e is in many cases similar to that in free solution when the dimensions of substrate molecules are much smaller than the pore diameter. Catalyst parameters and conditions associated with typical systems are summarized in Table 4, the corresponding effectiveness factors were estimated. Experimental effectiveness factors for immobilized cell systems as reported by Klein and Wagner are given in Table 5.

3.4.3 Complex Systems

Many enzymatic reactions involve *two substrates* or one substrate and a cofactor. In many cases the application of first order kinetics is a valuable approximation [22]. This is the case when the concentration of one substrate or coenzyme is sufficiently high and only the second substrate is rate-limiting. It applies for many analytical systems, e.g. for glucose (G) analysis by glucose oxidase in an air-saturated solution when $G < 10^{-4}$ (mol l^{-1}) (c.f. [52]). The same is true for many reactions which involve one substrate with low solubility (e.g. steroids, oxygen) when the other substrate is available in sufficiently high concentration. In these cases, the effectiveness (η) is not a function of the substrate concentration and conversion. η as a function of the Thiele modulus (φ) can be taken from Figs. 12 or 13 (where $S = 0.1K_M$).

In the intermediate concentration range (when both substrates are rate-limiting) η can exhibit a maximum [67]. This is shown in Fig. 14 for the glucose (G) oxidation by glucose oxidase (GOD) with oxygen as the second substrate. This phenomenon can be understood with the aid of calculated substrate profiles inside the catalyst matrix for different cases (Fig. 15). With a low glucose concentration in solution (left) the overall reaction is limited by the dropping profile of glucose concentration inside the catalyst matrix. At a high glucose concentration the overall reaction rate is higher (Fig. 14); however, the effectiveness is lower due to an even more steeply-dropping oxygen concentration. In the intermediate range, both substrate concentrations

Table 4. Calculated stationary (η) and operational (η_0) effectiveness factors for technical and analytical systems with immobilized enzymes [66] and from Figs. 12 and 13

System	$\frac{V_{max}}{K_M P_E}$	D_e $(cm^2\ s^{-1})$	d_p (μm)	$\left(\frac{S}{K_M}\right)^a$	η^b	$\eta_0{}^b$	Ref.
Immobilized Systems							
Trypsin-Sepharose	100	5×10^{-6}	100	100	1^c	0.7^c	[66]
Penicillin-acylase	100	5×10^{-6}	500	100	0.2	0.13	
				10	0.08	0.06	
Glucose-oxidase	100	5×10^{-6}	500	1	0.06	0.02	[67]
		5×10^{-6}	100	1	0.1	0.08	
Lactase	10	5×10^{-6}	500	100	0.65	0.45	[68]
		5×10^{-6}	500	10	0.17	0.12	

[a] S initial content under operating conditions.
[b] for Sh = 8 and (for η_0) 90% substrate conversion. No substrate or product inhibition.
[c] experimental values equal to calculated value within the experimental error.

Table 5. Experimental effectiveness for immobilized cell systems. a) variation of particle radius, b) variation of cell loading, c) variation of cell activity [28]
Candida tropicalis (Cast. Berkhout) immobilized in polyalcylamide; reaction: degradation of phenol T = 30 °C; pH = 6.7; air saturated oxygen concentration

	Radius R mm	Cell loading %	Free cell activity U per g wet weight	Catalyst activity U per ml cat.	Effectiveness
a)	0.125–0.2	15	10.3	1.2	0.78
	0.2–0.25	15	10.3	0.7	0.45
	0.25–0.395	15	10.3	0.68	0.44
b)	0.2–0.25	5	10.3	0.43	0.84
	0.2–0.25	15	10.3	0.7	0.45
	0.2–0.25	22	10.3	0.87	0.39
c)	cell strains with different levels of free cell activity for a given reaction.				

Example: *E. coli* cells in epoxy beads, 70% loading, particle radius R = 0.25 mm; reaction: formation of 6 APA from penicillin G (5%), 37 °C; pH = 7.8

Cell strain	Free cell activity U per g wet weight	Imm. cell act. U per ml cat.	Effectiveness
ATCC 11 105 (without T-shift)	15	3.3	0.32
ATCC 11 105 (without T-shift)	23	4.6	0.30
ATCC 11 105 (with T-shift)	61	9.8	0.24
E. coli 5K[a] (pH 12)	133	9.2	0.10

[a] Mayer, H., Collins, J., Wagner, F., Cloning of the penicillin-G acylase gene of *E. coli* ATCC 11 105 on multicopy plasmids. Enzyme Engineering Vol. 5, p. 61, New York: Plenum Press 1980

exhibit a pronounced profile, but less steep than in both limiting cases. Another example has been analyzed experimentally and by model calculations, leading to an effectiveness range from 0.05 to 0.8 [69].

A general theoretical treatment of two substrate kinetics and asymptotic approximations has been presented by Atkinson and Lester [70]. The limitations of such analysis very much depend on the accuracy of the experimental data. Thus, it is most difficult to estimate the amount of active immobilized enzymes when one substrate has a limiting solubility.

Immobilized systems of *two enzymes* catalysing a reaction sequence may exhibit enhanced efficiency. This has been shown theoretically for the nonstationary period of a reaction by Goldman and Katchalski [71] and experimentally by Mosbach and Mathiasson [72]. It is especially true for stationary reactions, as it has been shown experimentally [52, 73, 74] as well as theoretically [67, 75].

In the system of co-immobilized glucose oxidase and catalase, one product of glucose oxidation is H_2O_2 which is decomposed by catalase into H_2O and $1/2\ O_2$,

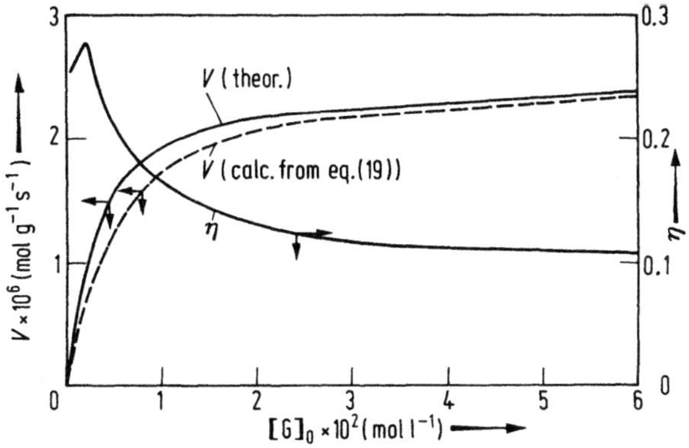

Fig. 14. Reaction rate (v) and effectiveness η as a function of one substrate (glucose G), where the second substrate (oxygen, saturated solution) has poor solubility. Glucose oxidation with immobilized glucose oxidase, v based on dry weight of carrier with water uptake of about 5 ml g^{-1}, $d_p \simeq 0.4$ mm. -------- calculated from pseudo-homogeneous approximation of experimental data [52]; ——— calculated from a model with coupled mass transfer and reaction [67]

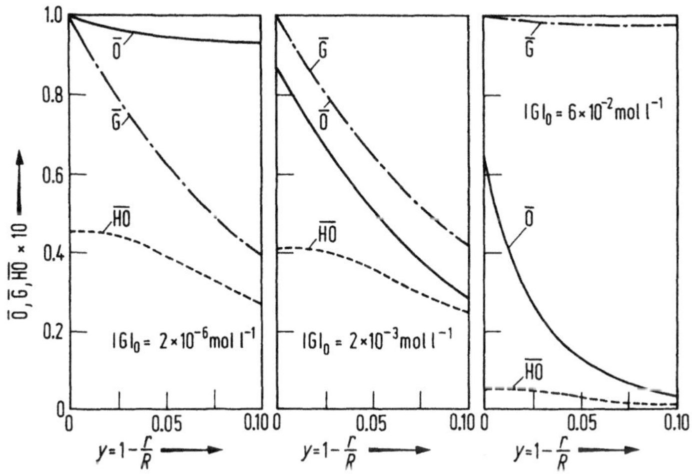

Fig. 15. Calculated profiles of substrates for the glucose oxidation inside a catalyst particle, normalized with respect to solution concentrations ($\bar{G}$: glucose, $\bar{O}$: oxygen; H_2O_2 as a product is also given: $\overline{HO}$, other data see Fig. 14). Abscissa: radial coordinate, up to 1/10 of particle radius (O: particle surface). Three examples for different glucose concentrations in solution at oxygen saturation are shown [67]

which in turn serves as a substrate for the first reaction. Owing to the proximity of both enzymes inside the carrier and to the restricted mass transfer — which leads to enhanced local concentrations of intermediate products — both enzymes exhibit higher reaction rates compared to their performance in single step reactions, or in a non-restricted environment. Thus glucose oxidation slows down by a factor of about 2 when catalase is inhibited. The H_2O_2 decomposition rate is slower by a factor of nearly 3 when H_2O_2 is supplied from the external solution, compared to the case where it is produced inside the carrier by glucose oxidase [52]. In the latter case the profile of the H_2O_2 concentration inside the matrix exhibits a maximum (Fig. 15) due to the coupled steps of production, decomposition and diffusion. Optimal ratios of activities for such immobilized enzyme systems can be found experimentally and theoretically [67, 74].

The effect of inhibition in immobilized systems has been discussed in [63]. It has been shown [76] that diffusional limitation in systems with substrate inhibition can lead to effectiveness factors >1. In such systems multiple steady states may occur [64] and criteria for such situations have been developed [77]. Further complications can arise with triphasic systems, when a reaction catalyzed by immobilized cells leads to gaseous products like methane. The gas may damage the catalyst or strongly restrict the transport of substrates in the porous carrier [78].

4 Operational Stability

A recent review [79] summarizes published investigations and suggests determination methods. Important parameters for both storage stability and operational stability are given.

Manifold phenomena are involved in inactivation of immobilized biocatalysts:
Physical parameters (pH, temperature etc.);
Chemical compounds (poisoning, toxic compounds);
Leakage of enzyme, attrition;
Microbial contamination;
Fouling, clogging.

It is important to consider, and to state, all parameters which are significant with respect to these phenomena:
Substrate and product concentration;
Concentration of solutes, especially toxic ones, metal ions etc.;
pH, buffer, ionic strength;
Temperature;
Time;
Mode of operation (column, stirred tank batch or continuous).

Due to many diverse factors, only the experimental determination of activity as a function of time (operational) leads to reliable results. Mechanisms underlying protein inactivation have been investigated and several aspects are basically understood. Thus one important mechanism is the oxidation of essential SH-groups [80]. However, the complex phenomena involved in the activity loss of immobilized biocatalysts can in general only be described empirically. First order kinetics have been applied most often, but basically no straight forward kinetics apply and extrapolation beyond experimental evidence may lead to errorneous results in many cases [28,79,81–83].

One of the most important parameters is temperature. In general high activation energies are involved in inactivation processes. Thus for immobilized invertase, 200–400 kJ per mol (50–95 kcal per mol) have been reported [81]. It has been found in several cases that the deactivation rate is a function of the substrate concentration [52,82,83]. This problem arises for many oxidation reactions, e.g. oxidation of amino acids to keto acids [82], of heterocyclic compounds [84], and of olefins [85,86]. One reaction product of oxidase-catalyzed reactions is H_2O_2, which inactivates proteins. Therefore the co-immobilization of catalase or an inorganic catalyst which decomposes H_2O_2 is necessary. It has been shown experimentally and theoretically that high substrate concentrations lead to high quasi-stationary H_2O_2 concentrations and in consequence to a rapid inactivation, which both increase approximately in linear correlation with the overall reaction rate. At low substrate concentrations, the enzyme system can exhibit long term stability [52,67] (Fig. 16). This is the case for analytical systems, e.g. for glucose determination [87], and for oxygen removal in wine or beer [88,89].

Another example where inactivation is coupled to the substrate concentration and the reaction rate is that of hydrolytic reactions, where the addition of alkali for neutralization causes a local pH-shift which inactivates the enzyme [3].

Several approaches have been developed for improving the catalyst stability. It has been shown that the addition of several compounds can be favorable during storage [90,92]. Thus the activity of alcohol dehydrogenase was considerably stabilized by glycerol at −196, −20 and 4 °C and by α-glycerophosphate at 4 °C and 30 °C [91]. Glycols and mercapto-compounds have been applied frequently for the protection or regeneration of activity ([80], for a summary see [90]).

The co-immobilization of mercapto-groups inside the carrier can improve the operational stability considerably [93]. Co-immobilization of catalytically inactive proteins such as albumin, or simply protein byproducts of enzyme production, may improve the stability [27,94]. Stabilization has also been achieved by modification [95], e.g. with dextrans and dextrins [96], by polyethyleneimines present during immobilization via glutaraldehyde [97] and by coating with albumin [98].

A sophisticated technique has been developed for the special case of hydrolases where an annulus of inert protein near the external surface of the carrier provided protection for the enzyme fixed inside the carrier matrix against denaturing due to pH gradients [99].

Table 6. *Operational stabilities* of several immobilized enzymes (further data see also [3,101], and for storage stability [79])

Enzyme (activity) (U per carrier)	Substrate (concentration) (mol per l)	T (°C)	Operation time (d)	Residual activity (%)	Ref.
Aminoacylase	Aminoacid	50	65	50	102)
Esterase	Cephalosporin (0,1)	40	30	55	103)
Penicillinacylase	Penicillin	37	>17	50	104)
Penicillinacylase	Penicillin (0,15)	37	(25 cycles)	75	93)
β-Galactosidase	Lactose (0,14)	40	56	90	105)
(820 U per g)[a]	Whey	40	~56	60	105)
β-Galactosidase (215 U per g alumina)	Lactose		>80		43)
β-Galactosidase (1230–360 U per g Duolite)	Whey ultrafiltrate	40	120	50	106)
β-Galactosidase[b]	Whey (cleared)	35	45	90	107)
	Milk	55	30	70	107)
Glucoseisomerase	Glucose	62	33	50	108)
Amino acid oxidase	2.4-Dinitro-phenylhydrazine	26	8	50	82)

[a] referring to dry chitosan.

[b] 25 g enzyme immobilized on 100 g carrier with 50% activity recovery (Plexazym Röhm)

Covalent attachment to carriers can stabilize enzymes [100], and also multiple covalent binding to carrier surfaces increases the thermal stability [56]. With regard to contamination, an established procedure is sanitation.

Examples of operational stabilities are collected in Table 6 where, however, not all of the information important for comparison is available.

5 Optimization

Optimization is understood as the control of reaction conditions in order to obtain optimal results under given boundary conditions. It involves biochemical engineering and economic parameters (see Tables 1 and 7) with respect to both catalyst design and reaction engineering. It comprises also scaling up as one important technique. Some approaches which concern selected parameters will be summarized from recent literature.

Some requirements of technical processes can be met by the selection of an appropriate catalyst carrier (for example with respect to its chemical and mechanical stability, its density compared to that of the solution) and by the choice of the reactor and hydrodynamic conditions. For the limitation of effectiveness by pore diffusion no simple straight-forward solution is available.

Table 7. Additional parameters important with respect to economics

Productivity: kg (product) per kg (catalyst, wet weight)[a]
Residence time of catalyst in the process (including regeneration, sanitation); residual activity
Quality of product (purity, other compounds)
Concentration of product

[a] For industrial processes, productivities in the range of 100–250 g product per kg catalyst were reported for penicillin acylase. For glucose isomerase it was estimated that about 1,95 million t of high fructose syrup were produced by 1.300 t of catalysts in 1980 (109). For β-galactosidase processing of 70 t whey or 13 t milk by 1 kg of catalyst with 90% and 70% residual activity, respectively, were reported (107)

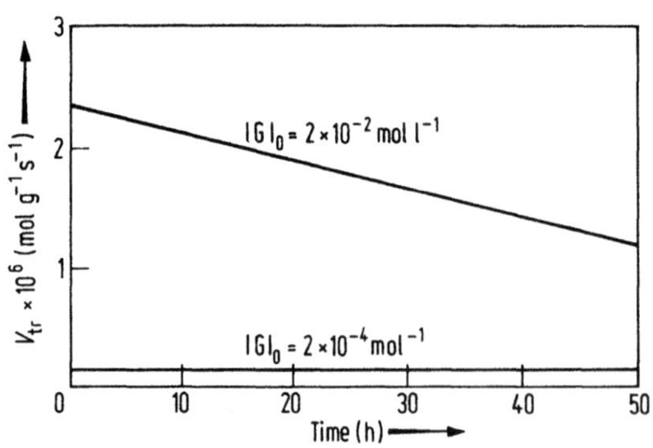

Fig. 16. Overall reaction rates of glucose oxidation by co-immobilized glucose oxidase and catalase as a function of operation time for two different substrate concentrations [67]

5.1 Optimization of Enzyme Distribution

Low catalyst effectiveness may be due to pore diffusion limitation associated with common sizes of particles and enzyme activities, and with either low substrate concentration or with hydrolytic reactions (where a high buffer concentration is unfavorable) or with substrates of high molar mass. In order to establish short diffusion paths one might select small carrier particles or carriers with an inert core and a thin porous shell carrying the enzyme. In general, neither solution can be applied because of technical and economic reasons.

Another solution takes advantage of the diffusion limitation during catalyst synthesis by immobilizing enzymes under conditions of kinetic diffusion control, resulting in a non-uniform distribution of enzyme throughout the matrix [110–113]. In general a uniform distribution of enzymes is assumed inside porous carriers [114]. Few investigations on this subject have been performed which show that radial gradients of enzyme density may exist due to an inhomogeneity of the matrix or to a kinetic control of immobilization [112–115]. An enzyme distribution such as that shown in Fig. 17 (taken from Carleysmith et al. [112]) provides short diffusion paths for substrates and products, thus reducing diffusion limitation of the overall reaction rate.

Several parameters are important for the synthesis of such biocatalysts, as can be shown by mathematical modelling: the enzyme concentration in solution (E_L), the ratio of the amount of enzyme and the binding capacity of the carrier, external mass transfer (Sh-number) during the synthesis, adsorption equilibrium (K_a), particle diameter (d_p), effective diffusion coefficient of the enzyme inside the porous matrix (D_e) and the rate of covalent coupling [116]. Figure 18 shows the calculated radial density profiles of adsorbed enzymes. They are most favorable for strong adsorption and at high Sh-numbers (high external mass transfer rate) [117]. High E_L result in steeper gradients for equal amounts of enzyme and carrier. Such profiles are favorable for particle diameters in the range of those most frequently used (0.1–1 mm). However, for an optimized enzyme distribution the external mass transfer may become the more important limiting factor for the catalyzed reaction. The kinetics of adsorption can be controlled by monitoring the decrease of enzyme concentration in free solution which can proceed in the range of a few minutes up to several hours.

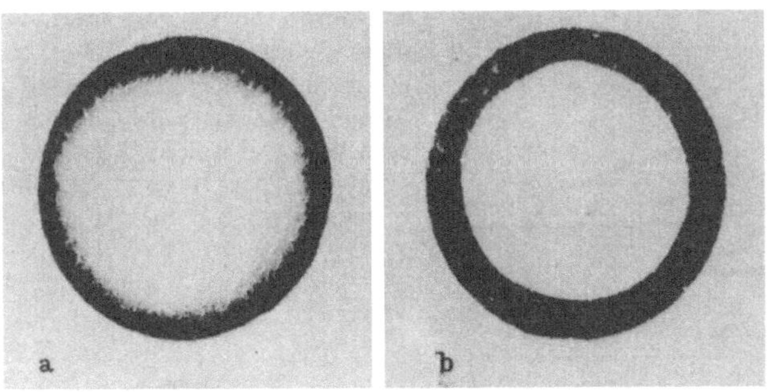

Fig. 17a and b. Photomicrographs of penicillin acylase (**a**) and bovine serum albumin (**b**) stained after immobilization on activated XAD beads; plan view of diametrically split beads. Dark stained zones show the location of protein [112]

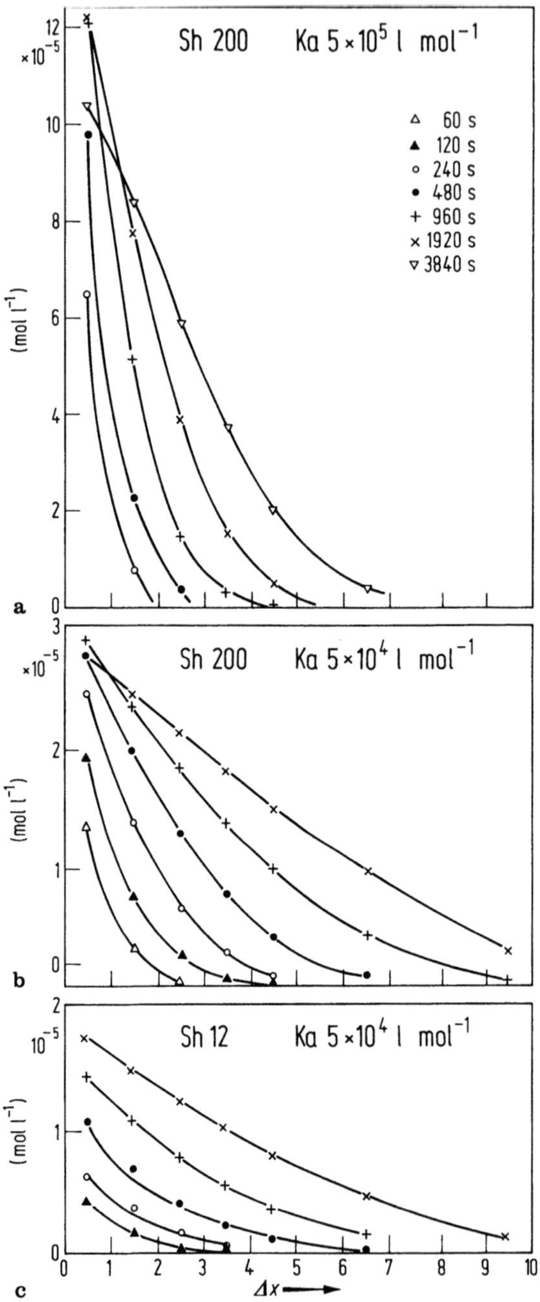

Fig. 18a—c. Calculated profiles of adsorbed enzyme density as a function of the distance (x) from the external carrier surface with adsorption time as parameter. **a** strong adsorption, equilibrium constant ($K_a = 5 \times 10^5$ (l/mol)) and high Sh-number (200), **b** weaker adsorption ($K_a = 5 \times 10^4$, Sh = 200), **c** low Sh-number ($K_a = 5 \times 10^4$, Sh = 12), adsorption times in the range of 60–1920 s are given. [Further parameters: E_L: $0.02\ g\,l^{-1}$, carrier: $20\ g\,l^{-1}$ with a capacity for enzyme $0.75\ g\,g^{-1}$, d_p: 1.5 mm, D_e: $5 \times 10^{-7}\ cm^2\,s^{-1}$, $\Delta x = 1$ corresponds to 0.01 mm [117]]

It has been shown experimentally that preparations with optimized enzyme distribution exhibit higher catalyst efficiencies when compared to those of equal amounts of enzyme immobilized without kinetic control, e.g. by 50% at low substrate concentrations, and by factors of up to 2.5 for high molar mass substrates [111,113]. This effect depends mostly on the parameters of the Thiele modulus, e.g. the total amount of immobilized enzyme. A maximum is found in the intermediate range between high and low loadings [117].

5.2 Further Concepts

Further concepts have been developed which are directed towards a specific design of the catalyst or the process in order to obtain maximum productivity (total amount of feed processed per unit of enzyme) for a given system. An example of an optimized catalyst design which takes inactivation into account has already been mentioned. It is based on the immobilization technique under conditions of kinetic diffusion control. Thus inert protein was bound only within 1/10 of the fractional penetration depth at the external catalyst surface, followed by immobilization of the enzyme penicillin acylase. The outer protein layer protected the enzyme from regions of elevated pH which resulted when alkali was added for product neutralization. The residual activity after repeated batch operation was raised to 93%, as compared to 68% for an unprotected preparation of similar overall activity [99].

Similarly maximum productivity can be obtained if oxidases are co-immobilized with catalase. Thus with a decreasing amount of immobilized oxidase its effectiveness as well as its operational stability will increase, due to the less severe influence of both the diffusion limitation of oxygen and inactivation by H_2O_2. For economic reasons, it is obvious that a minimum of catalyst activity must be maintained during operation. In consequence an optimum catalyst composition should exist in the intermediate range [67].

The catalytic reaction with immobilized enzymes may be influenced by specific properties of the catalyst surface [22]. Such perturbations can be used in order to shift the equilibrium of a reaction, e.g. toward synthesis of esters or peptides. Thus the synthesis of an ester by chymotrypsin, which otherwise catalyses its hydrolysis, could be observed when the enzyme was immobilized on a charged matrix [118]. This shift in equilibrium is due to dipole orientation energies at the carrier surface. The overall equilibrium of the system is controlled by the rapid enzyme-catalyzed reaction in the subsystem (the surface electric double layer).

In immobilized multienzyme systems, the activities of the enzymes involved should obviously be optimized. Thus an optimum exists for the ratio of glucose oxidase and catalase activities with respect to effectiveness [67], as well as for the ratio of invertase and glucose oxidase, when the yield of gluconic acid produced from sucrose is optimized [74]. This paper presents an optimum packing policy with respect to immobilized enzyme activities.

Reactor operation and control

Operating strategies were discussed by Vieth et al. [64] and recently by Pitcher [23]. Pitcher described an optimal policy for the control of substrate conversion by regulation of temperature and flow rate during continuous operation of a column reactor. Such a strategy was used in pilot plant units with immobilized lactase [119]. Constant productivity, in spite of decreasing catalyst activity, can also be maintained by means of a multiple reactor system.

Several parameters were investigated with immobilized penicillin acylase in a continuous four tank reactor system. pH-shift due to alkali addition was the main source of catalyst inactivation, which could be reduced by lowering the operational pH in the first tank, enhancing the stirring rate and diluting the alkali used for neutralization [104].

6 Conclusions

Much of the recent literature on immobilization has focussed on more complex systems, especially on immobilized cells. However, considerable progress has been achieved with the first generation systems which are now well established on an industrial scale. Progress in catalyst design and reaction engineering have succeeded in improving effectiveness and operational stability. Unfortunately rather little has been published on research and development performed in industrial laboratories. Thus only a small amount of information can be found on scale up concepts and, more general, information on important parameters of immobilized biocatalysts is poor in many publications.

It might be expected that improved techniques and economics will broaden the application range of simple immobilized enzyme systems. Thus reactions in organic solvents [120] or multiphase systems may open new routes for the modification of organic compounds. Simple and stable ready-to-use biocatalysts could find application in the degradation of toxic compounds in specific waste waters.

The progress reported in this article might also stimulate advances in more complex catalytic systems. Several techniques for cell immobilization have successfully been applied. This might favour the introduction of more complex systems on an industrial scale, and also the application of immobilized biocatalysts in the difficult field of medical application could witness progress in the near future.

7 Symbols

a, b, c		constants
D	$cm^2\ s^{-1}$	diffusion coefficient
D_e	$cm^2\ s^{-1}$	effective diffusion coefficient
d	cm, μm	diameter
E		enzyme
E	kJ per mol	activation energy
F	cm^2	surface
h	cm	height
I		ionic strength
K		equilibrium constant
K_M	$mol\ l^{-1}$	Michaelis constant

k	s^{-1}	kinetic rate constant (first order)
k_s	$cm\ s^{-1}$	mass transfer coefficient
L		length
n	s^{-1}	impeller rotation rate
P_E		fraction of accessible particle volume
Δp	Pa, bar	pressure drop
R	cm	radius
Re	$ud_p\nu^{-1}$	Reynolds number
Re_i	$nd_i^2\nu^{-1}$	impeller Reynolds number
r		radial coordinate
S	$mol\ l^{-1}$	substrate concentration
Sc	νD^{-1}	Schmidt number
Sh	$k_s d_p D^{-1}$	Sherwood number
T	°C	temperature
t	s, min	time
u	$cm\ s^{-1}$	flow rate (superficial velocity)
V	cm^3	volume
V_{max}	$mol\ l^{-1}\ s^{-1}$	maximum reaction rate
v_i		overall reaction rate
x		local coordinate

Greek Symbols:

δ		thickness of diffusion boundary layer
ε		voidage
η	$g\ cm^{-1}\ s^{-1}$	dynamic viscosity and effectiveness (dimensionless)
φ		Thiele modulus (c.f. Eqs. (8), (9))
ϱ	$g\ cm^{-3}$	density

Suffixes:

a	adsorption
c	referring to unit catalyst volume
i	impeller, or inhibition
I	inhibitor
L	bulk phase
p	particle
r	radial coordinate
R	external particle surface

8 Appendix

Recent publications have contributed new methods and improved understanding in heterogeneous biocatalysis.

For investigations on mechanical properties a method for testing the tensile strength has been developed [1], and experimental results have been given for carrageenan, agar and polyacrylamide carriers. Ca-alginate gels were found to be inhomogeneous with a stronger outer layer.

Many experimental results show decreasing activity vs. operation time where an abrupt change of the slope has been observed. Fig. A1 shows such results with epoxide-immobilized *E. coli* cells and the hydrolysis of penicillin G. Recent theoretical considerations by Klein et al. [2] on coupled

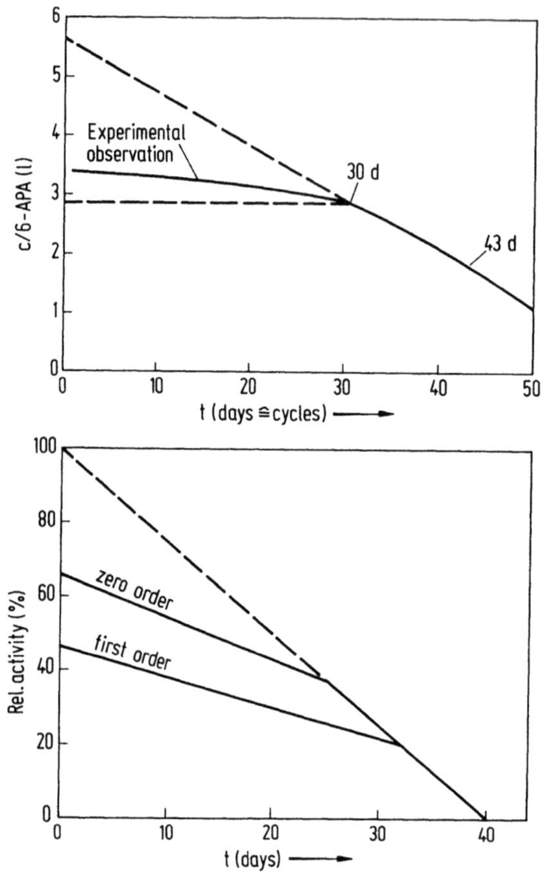

Fig. A1. Experimental results for decreasing activity with operation cycles (hydrolysis of 6-aminopenicillanic acid by immobilized *E. coli* cells (above). Model calculation with two different inactivation kinetics (below) [2]

diffusion, reaction and inactivation processes provide further insight into these complex phenomena. The slope observed initially is due to the diffusion limited overall reaction kinetics with superimposed inactivation. The reaction zone is gradually moving from the outer shell towards the center of the catalyst, presumably due to inactivation by alcali added for neutralization of the product (c.f. Sect. 5.1.2). The second phase with steeply decreasing overall activity is primarily due to cell inactivation, when the reaction zone is at the center of the catalyst particle. Fig. A1 shows model calculations with two different inactivation kinetics which show qualitative agreement with the experimental observations. A further theoretical model for first order inactivation kinetics and diffusion control of the overall reaction has been published including some experimental results [3].

A different situation has been observed for the inactivation of immobilized cells during repeated hydrocortisone transformations, however with somewhat similar apparent inactivation kinetics [4]. The half life was 140 days during which time 160 transformations were performed. During this period cells with substantially changed structure were found, and the number of lysed cells increased with time and also with the distance from the particle surface. This was assumed to be due to anaerobic conditions. After 6 months all cells inside the granules were lysed, except those located in the subsurface layers. Incubation in nutrient medium yielded a five fold increased activity and a somewhat higher increase in the number of viable cells. During continued operation cell lysis was observed with a similar pattern as described before.

Another method for studying inactivation patterns by means of radioactive labeled cell protein exhibited rather low endogenous proteolysis with *Escherichia alcalescens* [5].

9 Appendix References

1. Krouwel, P. G., Harder, A., Kossen, N. W. F.: Biotechnol. Lett. *4*, 103–108 (1982)
2. Klein, J., Vorlop, K. D., ACS Winter Sympos., Boulder, Colorado (1982), and Klein, J., Vorlop, K. D., Eng, H.: Poster, 2nd Eur. Congress Biotechnol., Eastbourne, GB, 1981
3. Ooshima, H., Harano, J.: Biotech. Bioeng. *23*, 1991 (1981)
4. Koshcheenko, K. A., Sukhodolskaya, G. V., Tyurin, V. S., Skryabin, G. K.: Eur. J. Appl. Microbiol. Biotechnol. *12*, 161–169 (1981)
5. Skodova, H., Chaloupka, J., Skoda, J.: Biotech. Bioeng. *23*, 2151 (1981)

10 References

1. Micheel, F., Evers, J.: Makromol. Chem. *3*, 200 (1949)
2. Manecke, G., Gillert, K. E.: Naturwiss. *42*, 212 (1955)
3. Lilly, M. D.: in Biotechnology, Dechema Monogr. *82*, 165 (1978)
4. Characterization of Immobilized Biocatalysts, Dechema Monogr. *84* (1979)
5. Buchholz, K. et al.: 4., 5a: p. 33; 5b: p. 34
6. Working Party on Immobilized Biocatalysts of the European Federation of Biotechnology, Poster, 2nd Eur. Congr. Biotechnology, Eastbourne GB 1981
7. Schlünsen, J., Ehrenthal, E., Manecke, G.: 4., p. 145
8. Klein, J., Wagner, F.: 4., p. 265
9. Immobilized Enzymes, Methods in Enzymology Vol. 44, Mosbach, K. (ed.), New York: Academic Press, 1976
10. Adv. Biochem. Engng., Vol. 10, Berlin: Springer 1978
11. Schlünsen, J., Ehrenthal, E., Manecke, G.: 4., p. 118
12. Zittan, L., Poulson, P. B., Hemmingsen, S. H.: Die Stärke *27*, 236 (1975)
13. Brauer, H., Mewes, D.: Chem.-Ing.-Techn. *44*, 93 (1972)
14. Norsker, O., Gibson, K., Zittan, L.: Starch/Stärke *31*, 13 (1979)
15. Buchholz, K., Gödelmann, B.: in: Enzyme Eng., Vol. 4, p. 89, New York: Plenum Press 1978
16. Klein, J., Washausen, P.: 4., p. 277
17. Vorlop, K. D., Klein, J.: Biotechnol. Lett. *3*, 9 (1981)
18a. Schlünsen, J., Manecke, G.: Angew. Makromol. Chem. *76/77*, 299 (1979)
18b. Manecke, G., Schlünsen, J., Ehrenthal, E.: 4., p. 81
19. Krämer, D. M.: 4., p. 88
20. Buchholz, K.: 4., p. 136
21. Klein, J., Eng, H.: 4., p. 292
22. Goldstein, L.: 9., p. 397
23. Pitcher, W. H.: 10., p. 1
24. Kasche, V., Buchholz, K.: 4., p. 208
25. Kula, M. R.: 4., p. 182
26. Ehrenthal, E., Schlünsen, J., Manecke, G.: 4., p. 188
27. Schmidt-Kastner, G.: Dechema-Jahrestag. 1977
28. Klein, J.: 4., p. 303
29. Klein, J.: Dechema Monogr. *82*, 142 (1980)
30. Chibata, I.: in Enzyme Eng. Vol. 4, p. 335, New York: Plenum Press 1978
31. Kobayashi, T., Laidler, K. J.: Biochim. Biophys. Acta *302*, 1 (1973)
32. Hamilton, B. K., Gardner, C. R., Colton, C. K.: in: Immobilized Enzymes in Food and Microbial Processes, p. 205, New York, 1974

33. Buchholz, K., Rüth, W.: Biotech. Bioeng. *18*, 95 (1976)
34. Ngo, T. T., Laidler, K. J.: Biochim. Biophys. Acta *525*, 93 (1978)
35. Klein, J., Hackel, U., Wagner, F.: ACS Symp. Series *106*, 101 (1979)
36. Mukherjee, R. N. et al.: Biotech. Bioeng. *22*, 543 (1980)
37. Karabelas, A. J., Wegner, T. H., Hanratty, T. J.: Chem. Eng. Sci. *26*, 1581 (1971)
38. Traher, A. D., Kittrell, J. R.: Biotech. Bioeng. *16*, 419 (1974)
39. Marazzo, W. A., Merson, R. L., Mc Coy, B. J.: ibid. *17*, 1515 (1975)
40. Toda, K.: ibid. *17*, 1729 (1975)
41. Adachi, S. et al.: ibid. *22*, 779 (1980)
42. Coughlin, R. W. et al.: Chem.-Ing.-Techn. *47*, 111 (1975)
43. Allen, B. R., Charles, M., Coughlin, R. W.: Biotech. Bioeng. *21*, 689 (1979)
44. Emery, A. N., Cardoso, J. P.: ibid. *20*, 1903 (1978)
45. Levins, D. M., Glastonbury, J. R.: Trans. Instn. Chem. Engrs. *50*, 132 (1972)
46. Calderbank, P. H., Moo Young, M. B.: Chem. Eng. Sci. *16*, 39 (1961)
47. Goldstein, L., Levin, J., Katchalski, E.: Biochemistry *3*, 1913 (1964)
48. Konecny, J., Slanicka, J.: Biochim. Biophys. Acta *403*, 573 (1975)
49. Manecke, G., Ehrenthal, E., Schlünsen, J.: 4., p. 193
50. Engasser, K. M., Horvath, C.: Biochim. Biophys. Acta *358*, 178 (1974)
51. Smith, J. M.: Chemical Engineering Kinetics, p. 428, New York: Mc-Graw-Hill 1970
52. Buchholz, K., Gödelmann, B.: Biotech. Bioeng. *20*, 1201 (1978)
53. Knights, R. J., Light, A.: Arch. Biochem. Biophys. *160*, 377 (1974)
54. Swaisgood, H. E., Horton, H. R.: 9., 504
55. Gabel, D., Kasche, V.: 9., 526
56. Koch-Schmidt, A. C., Mosbach, K.: Biochem. *16*, 2105 (1977)
57. Royer, G. P., Uy, R.: J. Biol. Chem. *248*, 2627 (1973)
58. Bourdillon, C., Barbotin, J. N., Thomas, D.: FEBS-Lett. *68*, 27 (1976)
59. Lasch, J., Koelsch, R.: Eur. J. Biochem. *82*, 181 (1978)
60. Buchholz, K., Duggal, S. K.: Poster, 2nd Eur. Congress on Biotechnology, Eastbourne, GB, 1981
61. Moo-Young, M., Kobayashi, T.: Can. J. Chem. Eng. *50*, 162 (1972)
62. Kobayashi, T., Laidler, K. J.: Biochim. Biophys. Acta *302*, 1 (1973)
63. Engasser, J. M., Horvath, C.: in: Applied Biochemistry and Bioengineering, Vol. 1, p. 129, Wingard, L. B., Katchalski-Katzir, E., Goldstein, L. (eds.), New York: Academic Press 1976
64. Vieth, W. R., Venkatasubramanian, K., Constantinides, A., Davidson, B.: loc. cit. p. 221
65. Kasche, V., Kapune, A., Schwegler, H.: Enz. Microb. Techn. *1*, 41 (1979)
66. Kasche, V.: 4., p. 224
67. Reuss, M., Buchholz, K.: Biotech. Bioeng. *21*, 2061 (1979)
68. Hudgins, R. P.: Can. J. Chem. Eng. *50*, 427 (1972)
69. Dahodwala, S. K., Humphrey, A. E., Weibel, M. K.: Biotech. Bioeng. *18*, 987 (1976)
70. Atkinson, B., Lester, D. E.: Biotech. Bioeng. *16*, 1321 (1974)
71. Goldman, R., Katchalski, E.: J. Theor. Biol. *32*, 243 (1971)
72. Mosbach, K., Mathiasson, B.: Acta Chem. Scand. *24*, 2093 (1970)
73. Koch-Schmidt, A., Mathiasson, B., Mosbach, K.: Eur. J. Biochem. *81*, 71 (1977)
74. Fernandes, P. M. et al.: Chemtech. 438, July 1975
75. Lin, S. H.: Biophysical Chemistry *7*, 229 (1977)
76. Kobayashi, T., Moo-Young, M.: Biotech. Bioeng. *15*, 47 (1973)
77. Kulkarni, B. D., Ramachandran, P. A.: Biotech. Bioeng. *22*, 1759 (1980)
78. Krouwel, P. G., Kossen, N. W. F.: Biotech. Bioeng. *22*, 681 (1980)
79. Palm, D.: 4., p. 245.
80. Couderc, R., Baratti, J.: Biotech. Bioeng. *22*, 1155 (1980)
81. Ooshima, H., Sakimoto, M., Harano, J.: Biotech. Bioeng. *22*, 2169 (1980)
82. Fink, D. J., Falb, R. D., Bean, M. K.: AIChE Sympos. Ser. *74*, 18 (1978)
83a. Do, D. D., Weiland, R. H.: Biotech. Bioeng. *22*, 1087 (1980)
83b. Sadana, A.: Biotechnol. Lett. *2*, 279 (1980)
84. Tramper, J., Müller, F., v. d. Plas, H.-C.: Biotech. Bioeng. *20*, 1507 (1978)
85. Cetus, preliminary prospectus, Jan. 1981

86. Hou, C. T., Patel, R. N., Laskin, A. J.: U.K. Patent Appl. 2.019.390 (1979)
87. Nelboeck, M., in: Biotechnological Application of Proteins and Enzymes, p. 279, New York: Academic Press 1977
88. Hartmeier, W.: Monatszeitschrift für Brauerei *31*, 16 (1978)
89. Schmid, R. D.: Process Biochem. *14*, No. 5, 2 (1979)
90. Schmid, R. D.: Adv. Biochem. Eng., Vol. 12, p. 41, Ghose, T. K., Fiechter, A., Blakebrough, N. (eds.), Berlin: Springer 1979
91. Brougham, M. J., Johnson, D. B.: Enz. Microb. Technol. *3*, 225 (1981)
92. D'Angiuro, L.: Biotech. Bioeng. *22*, 2251 (1980)
93. Krämer, D., Sauber, K.: German patent 2732301 (1980)
94. Kohlwey, D. E., Cheryan, M.: Enz. Microb. Technol. *3*, 64–68 (1981)
95. Tajima, M.: Eur. J. Biochem. *64*, 243 (1976)
96. Nakhapetyan, L. A., Akparov, V. Kh.: in: Enzyme Engineering Vol. 5, p. 423, Weetall, H. H., Royer, G. P. (eds.), New York: Plenum Press 1980
97. Gestrelius, S.: loc. cit., p. 439
98. Thomas, D.: Report from a visiting journey of German scientists to French research centres, 1980
99. Carleysmith, S. W., Eanes, M. B. L., Lilly, M. D.: Biotech. Bioeng. *22*, 957 (1980)
100. Cabral, J. M. S., Cardoso, J. P., Novais, J. M.: Enzy. Microb. Technol. *3*, 41 (1981)
101. Greenberg, N. A., Mahoney, R. R.: Process Biochem., p. 2, Feb./March (1981)
102. Chibata, J., Tosa, T.: Applied Biochem. and Bioeng. Vol. 1, p. 329, Wingard, L. B., Katchalski-Katzir, E., Goldstein, L. (eds.), New York: Academic Press 1976
103. Konecny, J., Sieber, M.: Biotech. Bioeng. *22*, 2013 (1980)
104. Carleysmith, S. W., Lilly, M. D.: Biotech. Bioeng. *21*, 1057 (1979)
105. Leuba, J. L., Widmer, F. J.: Solid Phase Biochem. *2*, 257 (1978)
106. Olson, A. C., Stanley, W. L.: J. Agr. Food Chem. *15*, 597 (1973)
107. Plainer, H., Sprössler, B. G.: Poster, 2nd Eur. Congress on Biotechnology, Eastbourne, GB, 1981
108. Fullbrook, P., Vabo, B.: J. Chem. E. Symp. No. 51, p. 31 (1977)
109. Poulson, P. B.: Enz. Microb. Technol. *3*, 271 (1981)
110. Buchholz, K., Borchert, A.: German patent P. 2805366.8 (1978)
111. Borchert, A., Buchholz, K.: Biotechnol. Lett. *1*, 15 (1979)
112. Carleysmith, S. W., Dunnil, P., Lilly, M. D.: Biotech. Bioeng. *22*, 735 (1980)
113. Guisan, J. M., Fernandez, V. M., Ballesteros, A.: in: Enzyme Eng. Vol. 5, p. 435, Weetall, H. H., Royer, G. P. (eds.), (1980)
114. Lasch, J. et al.: Eur. J. Biochem. *60*, 163 (1975)
115. Sernetz, M., Hannibal-Friedrich, O., Chun, M.: Microscopia Acta *81* (1979) (5)
116. Buchholz, K.: Biotechnol. Lett. *1*, 451 (1979)
117. Borchert, A.: Dissertation, Technical University of Braunschweig 1980
118. Kapune, A., Kasche, V.: BBRC *80*, 955 (1978)
119. Dohan, L. A. et al.: in: Enzyme Eng. Vol. 5, p. 279, Weetall, H. H., Royer, G. P. (eds.), New York: Plenum Press 1980
120. Fukui, S., Omota, T., Yamane, T., Tanaka, A.: in: Enzyme Eng. Vol. 5, p. 347, Weetall, H. H., Royer, G. P. (eds.), New York: Plenum Press 1980

Purification of Enzymes by Liquid-Liquid Extraction

Maria-Regina Kula, Karl Heinz Kroner, and Helmut Hustedt
Gesellschaft für Biotechnologische Forschung mbH
D-3300 Braunschweig-Stöckheim, Mascheroder Weg 1, FRG

The article reviews the current status of the application of aqueous two-phase systems for the extractive purification of enzymes, especially with regard to large-scale processing. The method can be used for the separation of proteins from cell debris as well as for further purification. The latter can be performed by a series of single step partitions, and apparently also by continuous multistage processes. The specificity and selectivity of extraction can be enhanced by introducing specific or general ligands. Scale-up of extractive enzyme purification is relatively simple utilizing commercially available equipment and machinery common in the chemical industry. Besides the technical performance, economic considerations also indicate the feasibility of the method at production scale.

1 Introduction

Filtration and centrifugation, the separation technologies presently used in industrial enzyme recovery and initial isolation processes are strongly dependent directly or indirectly on the particle size [1]. This may already present problems during harvesting of small bacteria, but the problems are magnified once the microorganisms are broken. After disintegration the size is even smaller while the viscosity of the suspension increases [2,3]. Both effects are detrimental for subsequent solid-liquid separation. On the other hand recovery processes for intracellular, biologically active proteins are of increasing importance. Future development of enzyme technology depends on the availability of intracellular enzymes at reasonable cost. Furthermore, production of certain proteins, e.g. mammalian proteohormones, in bacteria or yeast by recombinant DNA technology requires an adequate separation technology. There are several alternatives to improve the situation, e.g. flocculation to enlarge the particle size, use of higher g-forces to shorten settling times, use of filter aids to avoid clogging and compression of the filter cakes [1]. All these improvements however still keep the basic principle of solid-liquid separation. In contrast, we have been looking for some years now into extraction techniques for enzyme isolation and purification which are based on liquid-liquid separation [4,5]. Extraction is a highly advanced technology in other fields of the pharmaceutical and chemical industry and thought to be particularly useful when dealing with labile substances. However, the commonly used organic solvent systems are unsuitable for the intended purpose as proteins in general are either insoluble or become denatured in organic solvents. But phase separation can also be brought about by adding polymers to a solution. The phenomenon is called "incompatibility of polymers" and has been known for aqueous two-phase systems since 1896, when Beijerinck [6,7], described the formation of two liquid phases while mixing agar with soluble starch or gelatine. Today other hydrophilic polymers are used giving rise to two (or more) immiscible aqueous phases if the concentration of the polymers exceeds certain threshold levels. Aqueous two-phase systems contain a high proportion of water in both phases and can provide an excellent environment for cells, cell organelles or biologically active proteins [8]. Such systems appeared quite promising for extraction, capable of suspending cells and cell debris and partitioning the desired products and cell debris or other substances to be removed in opposite phases. In the following chapters biochemical and engineering aspects of partition and phase separation will be discussed in detail. The initial extraction of enzymes for the removal of cell debris was soon followed by additional partition steps for further purification of enzymes [9,10], especially for use as industrial catalysts. In many cases, contaminating activities or undesirable byproducts, such as nucleic acids and polysaccharides, can be removed rapidly and effectively by extraction. Furthermore, several approaches are presently pursued to improve the specificity and selectivity of extraction [11,12]. So far, basic technology for extraction of biologically active proteins has been developed to pilot plant scale in our laboratory, handling 100–200 g enzyme protein or several million units of activity [13,14]. General economic aspects are evaluated in the last chapter. Industrial application appears feasible and is expected in the near future.

2 Properties of Aqueous Phase Systems

2.1 General Comments

The physical chemistry of polymers in solution underlying the so-called incompatibility is dealt with in text books of polymer science [15]. Aqueous multiple-phase systems were studied in detail by P. A. Albertsson and his collaborators for more than twenty years [8]. The basic principles involved in partition as well as experimental aspects for laboratory scale are treated in the monograph "Partition of Cell Particles and Macromolecules", by P. A. Albertsson which is recommended as a thorough introduction to the field. The following summary will necessarily be short and relies heavily on P. A. Albertsson's work.

2.2 Phase Diagrams

In principle most, if not all, of the hydrophilic, natural or synthetic polymers miscible with water will show phase separation in a mixture with a second polymer or with salts. Figures 1 and 2 show representative examples for such behaviour in commonly employed systems. In Fig. 1 the phase diagram of a polyethylene glycol-dextran system is given. Both polymers are separately miscible with water in all concentrations. Below the binodal curve a homogeneous solution will be obtained. But if certain concentrations are exceeded, phase separation takes place and a PEG-rich upper phase and dextran-rich lower phase are formed which are no longer miscible with each other, despite the fact that both phases contain a high proportion (>75%) of water.

In Fig. 2a phase diagram for a polyethylene glycol-potassium phosphate system is presented. The tie-lines connecting phase compositions on the binodal curve which are in equilibrium with each other, are steeper than in polyethylene glycol-dextran systems. All mixtures with compositions represented by points on the same tie-line from T through M to B will yield phases with identical compositions of top and bottom phase but different phase volumes. Systems on the same tie-line therefore

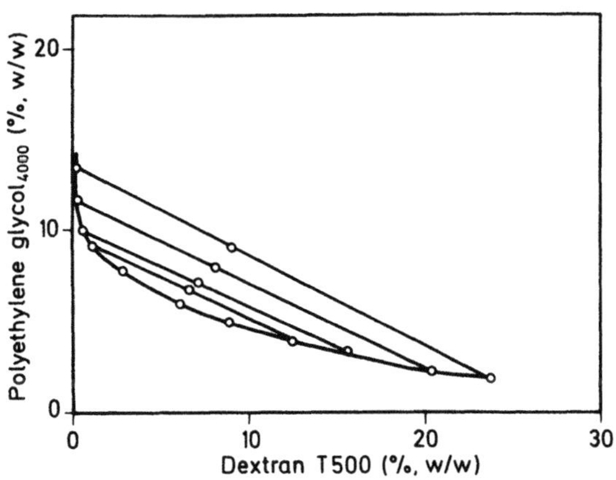

Fig. 1. Phase diagram of the system polyethylene glycol 4000/dextran T-500 at 20 °C (data from Albertsson [8])

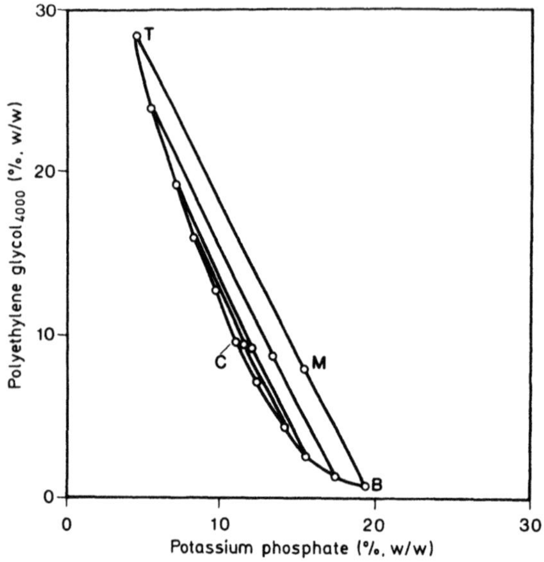

Fig. 2. Phase diagram of the system polyethylene glycol 4000/potassium phosphate at 20 °C (data from Albertsson [8]).
C = critical point;
T = composition of the top phase;
B = composition of the bottom phase;
M = composition of the total system

exhibit identical partition coefficients. This simplified treatment, however, does not take into account influences of the polydispersity of the polymers employed, which leads to some degree of polymer fractionation in aqueous phase systems and may therefore give rise to nonidentical partition coefficients for widely separated points on the same tie-line.

At the critical point, C, where the addition of a minute amount of water transforms a two-phase system to a homogeneous solution, the two phases should theoretically have identical compositions and volumes and give partition coefficients of 1.

2.3 Partition Coefficients

The partition coefficient is defined by Eq. (1).

$$K = \frac{c_T}{c_B} \tag{1}$$

where c_T and c_B are the equilibrium concentrations of the partitioned compound in the top and bottom phases, respectively. The partition coefficient of enzymes is constant for a given system over a fairly wide range of concentration, provided no association or dissociation of oligomeric proteins takes place in one of the phases [8].

Any molecule will accumulate in the phase where maximum number of interactions are possible and partition in such way, that a minimum of the energy content of the system is reached. The Brønstedt equation [16,17]:

$$K = e^{\frac{\lambda M}{kT}} \tag{2}$$

where λ is a parameter characterizing the phase system and interactions with the compound of interest, M the molecular weight, k the Boltzmann constant and T the absolute temperature, describes qualitatively — λ-values are unknown — an exponential dependence of the partition coefficient on the molecular weight and the factor λ. If dealing with compounds possessing very high molecular weights, e.g. $M_r > 10^6$ dalton, a one-sided partition can be expected and small variations in λ can produce drastic changes of the partition coefficient.

The partition of proteins and other compounds in aqueous two-phase systems is influenced by a large number of parameters. The important variables are summarized in Table 1. Most, if not all of the parameters listed, do not act independently, therefore calculation or theoretical predictions of the partition coefficient for a given protein cannot be carried out at present. Suitable conditions for a desired partition have to be found by experimentation. This is aided considerably if a fast analytical determination for the compound of interest is available. The experimental reproducibility of partition coefficients is normally in the range of $\pm 5\%$ for any volume analyzed.

Table 1. Parameters influencing partition in aqueous two-phase systems

- types of polymers composing the two-phase systems
- average molecular weight of the polymers
- molecular weight distribution of polymers
- length of the tie-line (complex function of concentration)
- types of ions composing or added to the system
- ionic strength
- pH
- temperature

2.4 Polymers Constituting the Phase Systems

P. Å. Albertsson published phase diagrams for a number of different hydrophilic polymers [8], e.g. dextran/polyethylene glycol, dextran/ficoll, ficoll/polyethylene glycol, polyvinyl alcohol/dextran, dextran/methyl cellulose, and polyethylene glycol/salt. Development of large scale extractions are limited so far to systems made up with polyethylene glycol and dextran or with polyethylene glycol and salts. Besides their general applicability and relatively desirable physical properties, especially regarding the viscosity and density difference of the resulting phase systems, the choice of the polymers was influenced very much by regulatory requirements for production processes. Dextran as well as polyethylene glycol are nontoxic substances and have been thoroughly tested for pharmaceutical purposes. Both polymers are included in the pharmacopoeia of most countries and are also registered for food purposes. This was considered of considerable advantage when developing the new technology, since most applications for enzymes and biologically active proteins were initially in the pharmaceutical and food industries. There is no reason otherwise why other hydrophilic polymers may not be used with advantage to establish an aqueous phase system for the extraction of cell components.

For industrial application the use of highly purified dextran fractions is normally too expensive. A crude dextran of high molecular weight ($\bar{M}_w > 5 \times 10^6$) and a crude dextran after limited hydrolysis ($\bar{M}_w > 4 \times 10^5$) with dilute hydrochloric acid as phase forming polymers [18] were therefore evaluated as economic substitutes. Binodal curves arising from these products are given in Fig. 3 and compared with the T-500 fraction of dextran ($M_w = 3.6 \times 10^5$), previously used. The critical concentration to bring about phase separation decreases with increasing molecular weight of the dextran samples. The binodal curve for the hydrolyzed dextran is found lower than expected compared to the T-500 fraction [18]. In Table 2 the polymer properties of the three dextran samples are summarized. The values for $\bar{M}_w$ and $\bar{M}_n$ indicate that the average molecular weight is lower after hydrolysis, but the polydispersity is rather high and a mixture of large and small molecules is present in the hydrolized

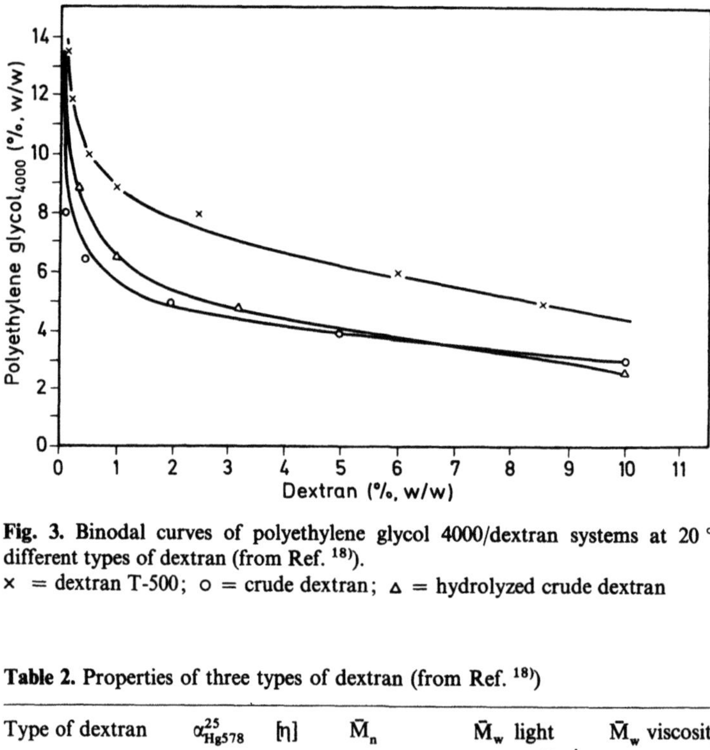

Fig. 3. Binodal curves of polyethylene glycol 4000/dextran systems at 20 °C, comparing three different types of dextran (from Ref. [18]).
× = dextran T-500; o = crude dextran; △ = hydrolyzed crude dextran

Table 2. Properties of three types of dextran (from Ref. [18])

Type of dextran	α^{25}_{Hg578}	$[\eta]$	$\bar{M}_n$	$\bar{M}_w$ light scattering	$\bar{M}_w$ viscosity[a]	$\bar{M}_w/\bar{M}_n$
I Crude dextran[b]	+177	0.86	13200	$\sim 5 \times 10^7$	5×10^6	$3.8–38 \times 10^2$
II Crude dextran hydrolyzed	+163	0.57	1200	3.6×10^5	4.4×10^5	$3.0–3.7 \times 10^2$
III T-500	+197	0.52	171000	3.7×10^5	3.5×10^5	2.1–2.2

[a] $\bar{M}_w$ calculated from viscosity measurements:
(I) graphically from a double logarithmic plot of $[\eta]$ versus $\bar{M}_w$
(II) and (III) using $[\eta] = 2.43 \times 10^{-3}\ \bar{M}_w^{0.42}$
[b] gift from Sorigona, A. B. Staffanstorp, Sweden

crude dextran samples. This will influence the binodal curve (Fig. 3) and eventually also the partition coefficient of proteins. Often an increase in the partition coefficient of proteins is observed with increasing molecular weight of dextran[8,19]. For a number of different enzymes partition experiments were carried out. Partition coefficients and yields are summarized in Table 3. Several phase systems are compared. The concentration of the polymers was selected to yield approximately the same length of tie-line in the resulting two-phase system using different dextran samples. The influence of the crude dextran on the partition coefficient was not as pronounced as expected. In all cases enzymes were recovered in high yield demonstrating that with regard to biochemical behaviour crude dextran as such or after mild hydrolysis can be used as the phase forming polymer. The influence of such changes on the

Table 3. Comparison of enzyme partition coefficients (from Ref.[18])

Enzyme	System	Volume ratio	K	ΔK	Yield (%)
α-amylase	I	8.16	3.64	—	96.7
	II	8.78	4.18	+0.54	97.8
	III	8.60	3.68	+0.04	97.2
Glucoamylase	I	8.60	1.22	—	91.3
	II	10.12	1.80	+0.58	95.0
	III	9.28	1.34	+0.12	92.6
α-glucosidase	I	6.31	2.84	—	94.7
	II	10.88	3.37	+0.53	97.4
	III	9.67	2.67	−0.17	96.3
Glucose-6-phosphate-dehydrogenase	I	8.20	0.62	—	84.9
	II	10.00	1.31	+0.69	92.9
	III	8.90	0.89	+0.17	89.9
Formate dehydrogenase	I[a]	8.30	0.94	—	88.6
	II[b]	8.00	1.36	+0.42	91.6
	III[b]	7.90	1.07	+0.13	89.4
Formaldehyde dehydrogenase	I[a]	8.30	1.17	—	90.6
	II[b]	8.00	1.29	+0.12	91.2
	III[b]	7.90	1.06	−0.11	89.3
Catalase	I[a]	8.30	1.53	—	92.6
	II[b]	8.00	1.29	−0.24	91.2
	III[b]	7.90	1.57	+0.04	92.5
Pullulanase	I[a]	8.8	2.10	—	94.9
	II[a]	7.9	1.78	−0.32	93.4
	III[a]	6.8	2.10	0.0	93.5

I = 9% w/w PEG 4000, 2% w/w Dextran T-500 systems including
II = 9% w/w PEG 4000, 1.25% w/w crude Dextran 0.3 M potassium
III = 9% w/w PEG 4000, 1.5% w/w crude Dextran hydrolyzed phosphate, pH 7.5

[a] including 0.05 M potassium phosphate, pH 7.5

[b] systems contained only 7% w/w PEG 4000 and 0.05 M potassium phosphate, pH 7.5

technological properties during separation and with regard to economic consequences are discussed below. However, it should be emphasized that the hydrolysis of crude dextran can easily be incorporated into the preparation of a dextran stock solution. The low concentrations of sodium chloride in the stock solution, introduced by the neutralization of the hydrochloric acid used for hydrolysis, will not normally disturb further applications.

Lowering the average molecular weight of polyethylene glycol is a strategy often employed to increase the partition coefficient for the protein of interest presumably by lowering the hydrophobicity of the polyethylene glycol-rich phase [8]. The molecular weight distribution of polyethylene glycol can also be easily manipulated by mixing standard fractions of different average molecular weights. This will change $\bar{M}_w$ as well as the molecular weight distribution $\frac{\bar{M}_w}{\bar{M}_n}$ and influence the phase diagram as discussed. Figure 4 illustrates an example where the partition coefficient of fumarase is shifted over 6 orders of magnitude by altering the relative proportions of polyethylene glycol 4000 and 400. At the same time the apparent partition coefficient for total protein is changed only 20fold from 0.3 to 6. The contributions from $\bar{M}_w$ and $\bar{M}_w/\bar{M}_n$ are not yet differentiated, but this approach is promising and valuable if contaminating activities have to be removed from an enzyme extract.

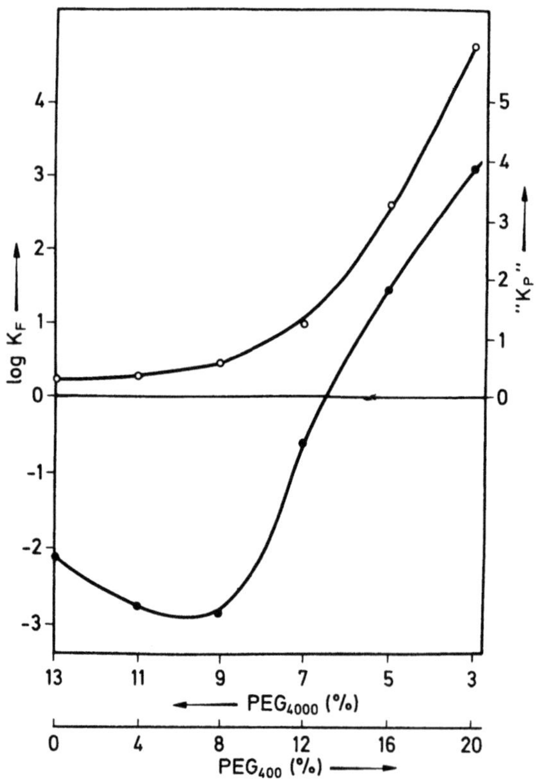

Fig. 4. Partition of fumarase and total protein as a function of the PEG 4000-PEG 400 content in a PEG/phosphate system including crude extract of *Brevibacterium ammoniagenes* (from Ref. [13]).
K_F = partition coefficient of fumarase (●)
"K_p" = apparent partition coefficient of total protein (○)

During extraction of enzymes from cell homogenates high concentrations of biopolymers e.g. cell debris, cell organelles, nucleic acids and proteins are included in the carrier system. Unter these conditions the concentration of biopolymers will be comparable to the concentration of the polymers constituting the phase system. This will lead to alterations of the phase system and may even result in complicated systems of higher order with multiple solid and liquid phases. Since polyethylene glycol is also a precipitating agent [20,21], partition in aqueous phase systems is limited by the solubility of the compound of interest. The dominating mechanism even in turbid systems can be identified by analyzing the concentration dependence, which is different for precipitation and partition. The concentration of the compound of interest will remain constant in the polyethylene glycol-rich phase for different starting concentrations if the phase is in equilibrium with a solid phase. In contrast, the concentration will increase in the top phases with the starting concentration if a partitionmechanism is operating. Precipitated material and cell debris has to be removed with the opposite phase for successful extraction of the desired protein.

2.5 Influence of Salt and Buffer Ions

Small ions act by several different mechanisms in aqueous phase systems. One important effect arises from the unequal partition of cations and anions constituting a salt or buffer in a phase system. This generates a small electric potential across the interface. Albertsson derived the following equations to describe the complex interactions for the partition of proteins in the presence of excess salt and the interfacial potential [8].

$$\ln K_p = \ln K_p^0 + (Z_p F/RT)\,\psi \qquad (3)$$

$$\psi = [RT/(Z^+ + Z^-)\,F]\,\ln(K_-/K_+) \qquad (4)$$

where ψ is the interfacial potential, Z_p the net charge of the protein, Z^+ and Z^- are the net charges of the cations and anions, K_p is the partition coefficient of the protein, K_+, K_- the partition coefficient of the cations and anions, respectively. R is the gas constant and F the Faraday constant, T the absolute temperature and K_p^0 is the partition coefficient of the protein in the system at zero interfacial potential or at the isoelectric point of the protein where Z_p becomes zero. For changes in the interfacial potential the ratio of ions rather than their concentration is decisive. For multivalent ions the dissociations is pH dependent and this in turn affects the partition coefficient of proteins. This interaction is very pronounced and often exploited in the case of phosphate buffers, where a large interfacial potential is induced above pH 7 by the unusually low partition coefficient of the HPO_4^{--} anion, which is only 0.74 in a 7% PEG 4000/7% dextran T-500 system [22]. This results in a negative charge into the lower phase and leads to a shift of negatively charged proteins into the upper phase. The unusual partition coefficient for the twice negatively charged HPO_4^{--} anion as compared to the $H_2PO_4^-$ anion (K = 0.96) carrying a single charge is attributed to a complex formation bridging regulary spaced hydroxyl groups in the polyglucan backbone of dextran by phosphate groups [23]. We have presented evidence that the gel formation resulting from this effect under certain conditions is accelerated in the presence of small amounts of polyethylene glycol and depends on the phosphate concentration and the pH [18]. The association of dextran chains will increase the exclusion volume above the increment corresponding to the single chains which in turn should lead to higher partition coefficients for any enzyme of a given size. This observation offers an explanation for the reported influence of the phosphate concentration on the partition of proteins [8,9,19], which is not due to a change in the interfacial potential. In Eqs. (3) and (4) no term is included with regard to the concentration of ions. The concentration of ions is only important in relation to the Donnan potential at the interface and is expected to be of no further influence above 20–50 mM concentration [8]. Changes in the partition coefficient of proteins at phosphate concentrations in the range 50–500 mM at pH-values above 7 reflect changes in K_p^0 due to steric exclusion. Addition of other divalent anions like SO_4^{--} will also induce gel formation and increase the partition coefficient of proteins but the effect is not as pronounced as with phosphate [18].

In polyethylene glycol/salt systems in addition, salting out effects appear to operate with increasing length of the tie-line shifting proteins from the salt phase into the PEG-rich phase or, if the solubility in the PEG-rich phase is not high enough, lead to protein precipitation at the interface. Figures 5 and 6 illustrate the behaviour observed with pullulanase from *K. pneumoniae*, but a quantitative description differentiating between several possible contributing forces cannot be given at present. Influence of different ions and their concentration should be investigated if a solution for a particular separation problem is sought. Solubility and salting out limits are individual properties of proteins, therefore a differential response is expected when

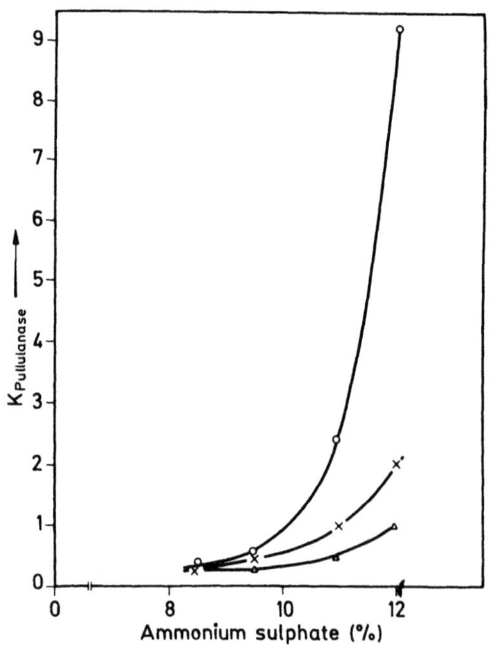

Fig. 5. Dependence of the partition coefficient of pullulanase on the total concentration of ammonium sulphate, in a system polyethylene glycol 4000/ammonium sulphate.
○ = 18% PEG 4000;
× = 16% PEG 4000;
△ = 14% PEG 4000;
(temperature: 20 °C)

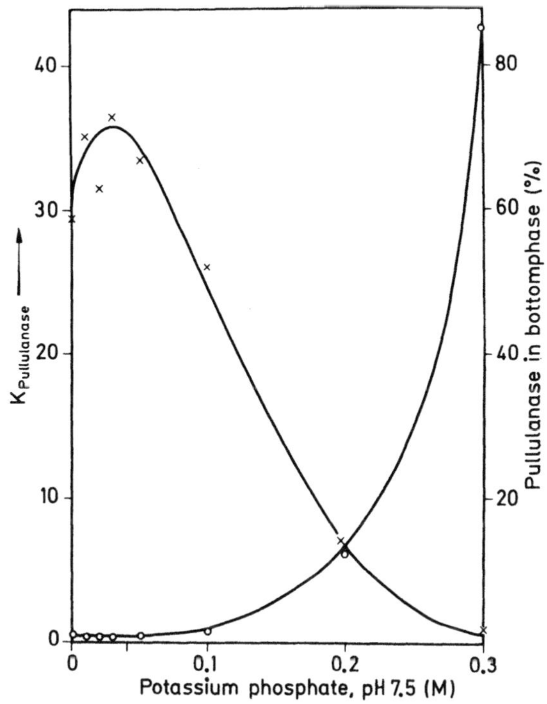

Fig. 6. Dependence of the partition coefficient of pullulanase on the concentration of potassium phosphate in a polyethylene glycol 4000/ammonium sulphate system.
System conditions: 14% PEG 4000, 9.5% $(NH_4)_2SO_4$.
o = partition coefficient of pullulanase;
× = yield of enzyme in salt phase;
(temperature: 20 °C)

a mixture of proteins is handled. Several examples are described in the literature, where such effects are utilized to improve purification and remove interfering activities: e.g. separation of glucosyl transferase and phosphorylase [9], purification of formate dehydrogenase [14] and the purification of fumarase [13].

3 Improvements in the Selectivity of Extraction

3.1 Basic Principles

The general parameters of a phase system listed in Table 1 are acting on common physico-chemical properties of proteins. Their differential separation power should not be underestimated (compare Fig. 11) but it is difficult to predict and to analyze. The finally observed partition coefficient of a protein is an integral over many possible interactions. P. Å. Albertsson [24] considers the following terms contributing to the macroscopic constant:

$$\ln K = \ln K_{el} + \ln K_{hphob} + \ln K_{hphil} + \ln K_{conf} + \ln K_{lig} \qquad (5)$$

where K_{el}, K_{hphob}, K_{hphil}, K_{conf} and K_{lig} denote partition coefficient increments due to electric, hydrophobic and hydrophilic forces or depending on the conformation or ligand interaction respectively. The state of hydration may be considered as an additional parameter in polyethylene glycol/salt systems. Alterations of parameters in commonly used phase systems most likely affect the first three terms in a complex fashion. However, it is also possible to manipulate these parameters more or less separately by including polymer derivatives in phase systems carrying appropriate groups. As long as the concentration of the polymer derivative is small compared to the unsubstituted species, it may be assumed that the polymer will direct the partition, so that the derivative behaves almost like the unsubstituted polymer, thereby accumulating the ligand predominantly into one phase. This approach necessitates the synthesis of various polymer derivatives. Some liquid ion exchangers based on dextran are commercially available, e.g. DEAE-dextran and dextran sulfate. Some polyethylene glycols with hydrophobic acyl groups are sold as surfactants [25]. However, the majority of interesting phase components still have to be prepared by the investigator, therefore progress is slow in this field. Polyethylene glycol derivatives are of special interest as proteins tend to avoid the polyethylene glycol-rich phase in PEG/dextran systems at low or moderate salt concentrations [8]. Any specific interaction shifting the desired protein into the PEG-rich phase is therefore expected to lead to a substantial purification. Substitutions on the polyethylene glycol are restricted to the terminal hydroxyl groups. Recently some new derivatives for "general ligand" affinity partition [26,27], hydrophobic partition [28] and partition by liquid ion exchangers [29] were prepared in our laboratories, and the chemical synthesis of polyethylene glycol intermediates was improved significantly in an attempt to make such modified polyethylene glycols available in larger quantities and at lower cost [27]. The application of some of these polyethylene glycol derivatives is discussed below.

3.2 Affinity Partition

As in conventional affinity chromatography biospecific interactions of enzymes with substrates, products, inhibitors or antibodies appear to offer the most rational design for binding and separation of a single protein from a complex mixture. Such a process was called affinity partition by Flanagan and Barondes [30]. It relies on the influence of the term K_{lig} in Eq. (5) on the partition coefficient. Applications have been discussed, e.g. by Hubert et al. [31] for the isolation of a 3-oxosteroid isomerase and by Takerkart et al. [32] for the isolation of trypsin. A multistage process for affinity partition and the recovery of modified liquid polymer is somewhat more difficult to carry out compared to conventional affinity chromatography using solid matrices. There are, however, specific advantages to a partition process. The binding capacity per unit volume seems to be considerably higher in aqueous phase systems [12] since the ligand density and availability can be increased. In addition, approach to equilibrium binding is faster in solution. In principle partition can be performed continuously in multistage operations, while chromatography is intrinsically a batch operation. Both aspects are important considerations for a large scale process. So despite the obvious disadvantages the concept of affinity partition cannot be dismissed. Developmental work is considerably less advanced than for simple extractions. Much more experimental data are needed before the potential of this technology can be estimated and the economic evaluations become more significant.

For reasons similar to those discussed in the literature concerning affinity chromatography, general ligands [33,34] appear especially useful reducing the otherwise necessary synthetic work load. With this in mind, the influence of PEG-NADH on the partition coefficient of some dehydrogenases was measured in a standard system. The results are shown in Figs. 7 and 8. The expected increase in the partition coefficient is clearly evident. It is related to the number of binding sites (n) in the oligomeric enzymes. The dimeric enzymes formate dehydrogenase and formaldehyde dehydrogenase exhibit by far the lowest observed partition coefficients, while glutamate dehydrogenase with 6 subunits has the highest. There appeared no to be quantitative correlation between the absolute values of the coefficients with known molecular parameters such as protein molecular weight, the number of binding sites or kinetic parameters K_m or K_i, indicating a more complex interaction. The partition coefficient in such selective systems is sensitive to the ionic strength, as can be seen from the influence of various ions on the partition of glucose-6-phosphate dehydrogenase in the presence of PEG-Cibacron-Blue (Fig. 9). The specific extraction is lost even at low to moderate ionic strength. This fact precludes the application of a PEG-salt system for such purposes. From these results, it also appears that the conditions employed for the comparison of the interaction between PEG-NADH and various dehydrogenases (Figs. 7 and 8) may have been suboptimal. These examples again serve to emphasize the intricate interdependence of numerous variables operating in aqueous phase systems, e.g. ionic strength, interfacial potential and binding equilibria. In the experiments described in Fig. 9 a triazine dye coupled to PEG was used as affinity agent. Triazine dyes covalently bound to hydrophilic resins are increasingly used for general ligand affinity chromatography [35,36]. Purification can be further improved by biospecific desorption employing a gradient of free coenzyme in buffer as eluent. We have shown that interactions of a dehydrogenase with PEG-

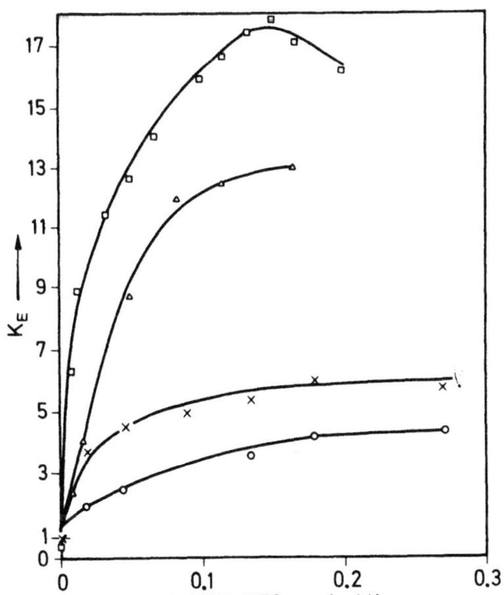

Fig. 7. Partition coefficient K_E of various dehydrogenases as a function of the concentration of PEG_{6000}-NADH in a PEG 4000/dextran T-500 system (data from Ref. [54])).
System conditions: 7% w/w PEG 4000, 6% w/w dextran T-500, 0.05 M potassium phosphate, pH 7.5, and 0.1 mM β-mercaptoethanol, temperature: 20 °C.
□ = lactate dehydrogenase (rabbit muscle);
△ = alcohol dehydrogenase (yeast);
× = formate dehydrogenase (*Candida boidinii*);
○ = formaldehyde dehydrogenase (*Candida boidinii*)

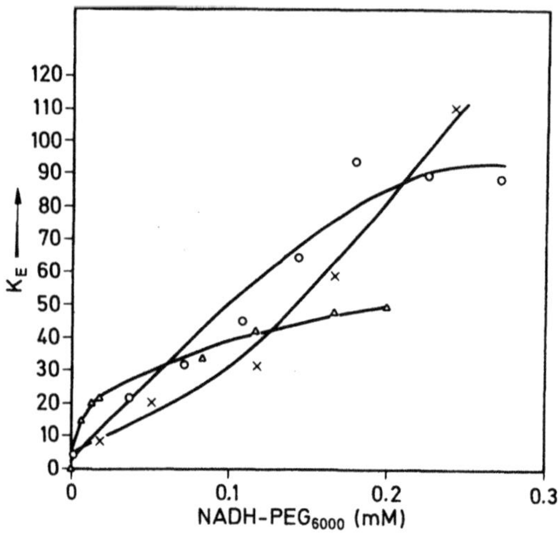

Fig. 8. The dependence of the partition coefficient K_E of alanine dehydrogenase, glutamate dehydrogenase and lactate dehydrogenase on the concentration of PEG_{6000}-NADH in a PEG 4000/dextran T-500 system (from ref. Ref. [54]).
System conditions: as in Fig. 7.
o = alanine dehydrogenase (*Bacillus subtilis*);
× = glutamate dehydrogenase (bovine liver);
△ = lactate dehydrogenase (pig heart)

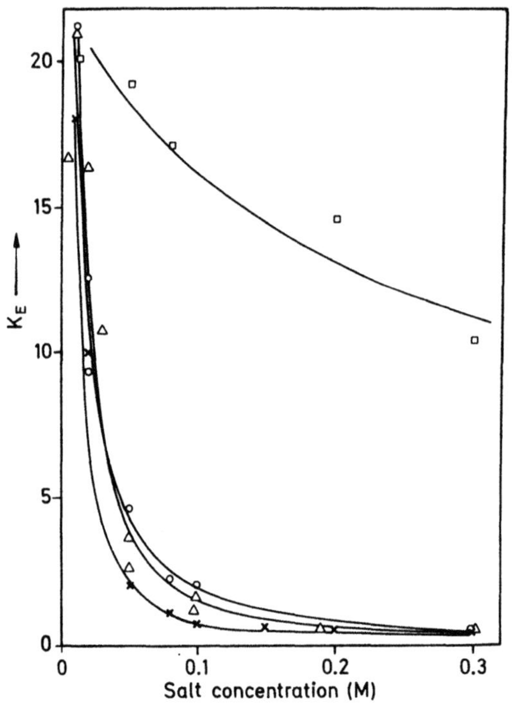

Fig. 9. Effect of different salts on the partition of glucose-6-phosphate dehydrogenase in the presence of PEG 6000-Blue (from Ref. [12]).
System conditions: 7% w/w PEG 4000, 5% w/w dextran T-500, 25 mg PEG 6000-Blue ad 5g total system.
△ = potassium phosphate, pH 7.5;
× = ammonium sulphate/TEA, pH 7.8;
o = potassium chloride/TEA, pH 7.8;
□ = sodium acetate, pH 7.5;
TEA = Triethanolamine buffer 5×10^{-3} M;
(temperature: 20 °C)

Cibacron-Blue in two-phase systems can be specifically suppressed by NADH [11]. Since triazine dyes can be bound to PEG without difficulties, these derivatives are cheap enough and could most likely serve for future practical applications. We therefore investigated the partition of glucose-6-phosphate dehydrogenase as a model enzyme in the presence of PEG-Blue and PEG-Red in detail [12].

As pointed out before, the capacity per unit volume is of considerable interest to evaluate affinity partition. The density of ligand groups in PEG-dextran systems is limited by the concentration and average molecular weight of polyethylene glycol, which determines the maximal number of endgroups available to bind the dye. It is questionable if very high, up to 100% replacement of the phase forming polymer with PEG-dye can be accomplished without excessively altering the overall properties of the system, but 50% substitution appears possible. Figure 10 illustrates the calculated binding capacity of such an affinity system for glucose-6-phosphate dehydrogenase. The capacity is indeed found to be remarkably high. The experimental data for PEG-Blue extrapolate to a value of 110 U per ml total system or 180 U per ml with

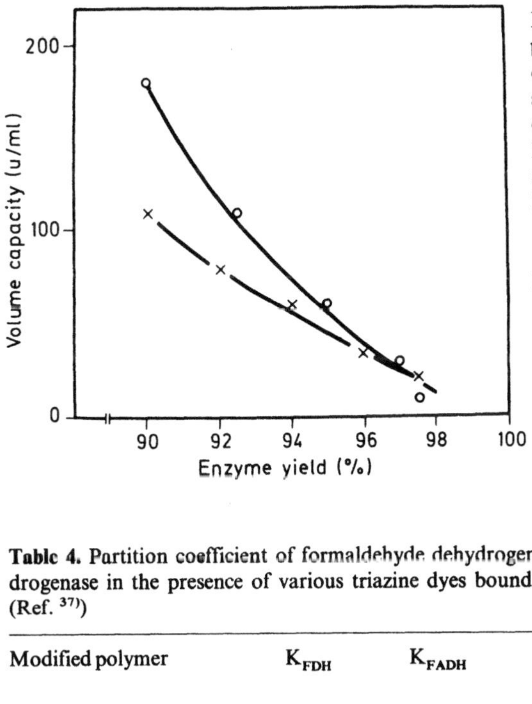

Fig. 10. Capacity-yield relation for top phases calculated for glucose-6-phosphate dehydrogenase from affinity partition studies with PEG 6000-Blue and PEG 6000-Red for a replacement of 50% of PEG in the carrier system by PEG-dyes. System conditions:
Blue — 7% w/w PEG 4000, 5% w/w dextran T-500, 0.02 M potassium phosphate, pH 7.8;
Red — 5.5% w/w PEG 6000, 7% dextran T-500, 0.05 M tris-acetate buffer, pH 7.8, 0.02 M $MgCl_2$, 0.4 mM EDTA.
o = PEG 6000-Blue (max. concentration ≃50 mg ml^{-1})
× = PEG 6000-Red (max. concentration ≃46 mg ml^{-1})

Table 4. Partition coefficient of formaldehyde dehydrogenase and formate dehydrogenase in the presence of various triazine dyes bound to polyethylene glycol (Ref. [37])

Modified polymer	K_{FDH}	K_{FADH}	Recovery	
			% FDH	% FADH
—	0.21	0.25	—	—
Procion red H-E7B	0.30	0.66	83	90
Procion red H-E3B	1.58	0.38	96	98
Procion green H-4G	0.26	0.34	80	94
Procion orange H-2R	0.24	0.64	87	103
Procion yellow 4-7G	0.23	0.27	90	100
Procion yellow MX-6G	0.35	0.46	nd	nd
Procion red MX-G	0.24	0.48	88	105
Procion red MX-5B	0.26	0.25	89	105
Procion green H-E4BD	0.33	0.32	95	97

System: 7% Dextran T-500
5% PEG 6000 (approximately 0.05% modified by the dye as indicated)
50 mmol per kg sodium acetate, pH 7.4
10% crude extract of *Candida boidinii*

Table 5. Recovery of Glucose-6-phosphate dehydrogenase after affinity partition with PEG_{6000}-Blue and PEG_{6000}-Red

Step	System composition (5 g total)	Total of PEG_{dye} mg	Initial units U	Total volume ml	Volume ratio V_T/V_B	Recovered units	Enzyme yield %	K_E	K_F
A	7 % PEG 4000 5 % Dextran T-500 0.02 M KP_i, pH 7.8	10 PEG-Blue	6.8	4.6	1.7	6.15 Top	90 Top	5.4	9.5
B	2 ml Top phase +12 % potassium phosphate, pH 7.8	—	4.2	2.05	0.5	4.30 Bottom	≃100 Bottom	<0.1	129
A	7 % PEG 4000 5 % Dextran T-500 0.05 M Tris-Ac, pH 7.8 0.02 % $MgCl_2$ 0.4 mM EDTA	5 PEG-Red	6.8	4.6	1.6	6.50 Top	95 Top	12.9	15
B	2 ml Top phase +12 % potassium phosphate, pH 7.8	—	4.6	2.05	0.41	4.6 Bottom	≃100 Bottom	<0.1	263

A = Affinity partition step
B = Re-partition step to a salt phase

respect to the upper phase for a 90% yield in a single extraction, when 50% of polyethylene glycol is replaced by the dye derivative [12].

Investigations of phase systems containing triazine dyes attached to polyethylene glycol as affinity agent are not completed yet. The first results [11,12,37] can only be regarded as preliminary. The type of dye is also important in the achievement of optimal results for a particular problem. This is illustrated in Table 4 summarizing changes in the partition coefficient of two quite similar enzymes, formate and formaldehyde dehydrogenase from *Candida boidinii* in response to various triazine dyes. The observed effects are relatively small, but apparently very selective. The strongest interaction was observed for polyethylene glycol-Red H-E3B and formate dehydrogenase. Desactivation of the enzymes by affinity partitioning was never observed in dye-containing systems. This can also be seen from data presented in Table 5 for glucose-6-phosphate dehydrogenase. After separation of the upper phase of an affinity system salt can be added to induce a secondary polyethylene glycol/salt system and the enzyme extracted into the salt phase. In this way, a major portion of the modified polyethylene glycol can also be recovered.

3.3 Liquid Ion Exchangers

Incorporation of liquid ion exchangers into aqueous phase systems can be envisaged as an analogous process to batch adsorption or column chromatography on solid ion exchanger resins. The use of liquid ion exchangers has a strong influence on the interfacial potential by confining multivalent ions to one phase [8]. This will be superimposed upon the unequal partition of small ions described in Eqs. (3) and (4). Detailed and systematic studies are still rare, but single experiments demonstrate that the principle can be applied [8,38]. Table 6 summarizes some results observed during an investigation of the partition of interferon β (human fibroblast interferon) by Menge et al. [29]. Here selective enhancement of the partition coefficient for interferon β versus total protein is observed by including α-(2-phosphoric acid monoethylester)-ω-phosphoric acid monoesterpoly(exyethylene) or α-(2-trimethylammonioethyl)-ω-trimethylammoniopoly(oxyethylene) into phase systems. One would expect that anion exchangers and cation exchangers would show opposite effects, yet both selectively extract interferon β into the polyethylene glycol-rich phase. The underlying mechanism presently remains obscure but seems to be related to hydrophobic interactions. This interpretation is strengthened by the observation that α-(2-phosphoric acid monoethylester)-ω-phosphoric acid monoesterpoly(oxyethylene) can also be applied in polyethylene glycol/salt systems, where ionic interactions should be negligible. Here the high discrimination between interferon β and contaminating protein is especially noteworthy (Table 6).

Table 6. Influence of different polyethylene glycol derivatives on the partition of Interferon (from Ref. [29])

Polyethylene glycol derivative	PEG derivative %(w/w)	Dextran T-500 %(w/w)	KH_2PO_4/ K_2HPO_4 %(w/w)	Ethylene glycol %(w/w)	NaCl ($mol\ l^{-1}$)	pH	Partition coefficient Interferon	Partition coefficient Protein
$(CH_3)_3N^+$—PEG-6000—$N^+(CH_3)_3$	8	8	—	—	1	6.0	4.70	0.41
$(CH_3)_3N^+$—PEG-6000—$N^+(CH_3)_3$	8	8	0.8	—	1	5.4	15.2	0.04
$(CH_3)_3N^+$—PEG-6000—$N^+(CH_3)_3$	8	8	0.8	25	1	4.6	49.5	0.12
$(CH_3)_3N^+$—PEG-1550—$N^+(CH_3)_3$	9.4	7.8	0.78	30	0.5	7	6.7	0.43
$(CH_3)_3N^+$—PEG-5000—OCH_3	8	8	0.8	—	0.5	7	3.8	0.011
NH_2—PEG-6000—NH_2	8	8	0.8	12	0.5	5	4.5	0.035
C_8H_{17}—NH_2^+—PEG-6000—NH_2^+—C_8H_{17}	8	8	0.8	—	1.5	4.8	20.7	n. d.
Cibacron blue—NH—PEG-6000—NH—Cibacron blue	8	8	0.8	—	0.5	6	>1	0.30
ConA—PEG-6000—ConA	8	8	0.8	—	0.5	6	37	0.05
HO_2C—$(CH_2)_2$—CO—NH—PEG-6000—NH—CO—$(CH_2)_2$—CO_2H	8	8	0.8	—	1	5.2	3.5	0.2
HO_2C—$(CH_2)_2$—CO—NH—PEG-6000—NH—CO—$(CH_2)_2$—CO_2H	8	8	0.8	—	1	6.9	2.86	0.09
HO_2C—$(CH_2)_2$—CO—NH—PEG-6000—NH—CO—$(CH_2)_2$—CO_2H	8	8	0.8	25	1	5.2	11.7	0.32
HO_2C—$(CH_2)_2$—CO—NH—PEG-6000—NH—CO—$(CH_2)_2$—CO_2H	8	8	0.8	25	1	6.6	30.2	0.09
H_2O_3P—O—PEG-6000—O—PO_3H_2	8	8	0.8	—	1	5.2	>133	0.08
H_2O_3P—O—PEG-6000—O—PO_3H_2	8	8	0.08	—	1	6.2	>109	0.05
H_2O_3P—O—PEG-6000—O—PO_3H_2	8	8	0.8	—	1	7.0	c. 93	0.05
H_2O_3P—O—PEG-6000—O—PO_3H_2	8	8	0.8	25	1	5.1	> 41	n. d.
H_2O_3P—O—PEG-6000—O—PO_3H_2	8	8	0.8	25	1	6.1	> 27	0.06
H_2O_3P—O—PEG-6000—O—PO_3H_2	8	8	0.8	25	1	7.1	19.4	0.06
H_2O_3P—O—PEG-6000—O—PO_3H_2	14	—	14	—	0.5	6.7	170	0.04
H_2O_3P—O—PEG-6000—O—PO_3H_2	1	—	19.5	—	0.5	6.7	169–321	0.23–0.35

4 Single Step Partitions

4.1 Removal of Cell Debris

The Brønstedt equation (2) predicts that for large entities like cells or cell particles and cell debris with very large apparent values of M, a onesided partition should occur, which is indeed frequently observed. Protein partition coefficients are often found in the range 0.1–10. To achieve the desired single step separation between enzymes and cell debris, conditions have to be established so that the enzyme(s) of interest and cell debris prefer opposite phases. But, at the same time, the partition coefficient of the enzyme and the volume ratio of the phases should be sufficient to extract the enzyme in high yield. Therefore, high values for the partition ratio, G, are needed for a one step extraction. G is described by equation (6).

$$G = K \frac{V_T}{V_B} \tag{6}$$

The theoretical yields of extraction are given by Eqs. (7) and (8) for the top and bottom phases, respectively, where V_T and V_B are the volumes of top and bottom phases.

$$Y_T(\%) = \frac{100}{1 + \frac{V_B}{V_T}\frac{1}{K}} \tag{7}$$

$$Y_B(\%) = \frac{100}{1 + \frac{V_T}{V_B}K} \tag{8}$$

It is sometimes complicated to obtain a clear extract if precipitates are present. So far, however, conditions could always be found to remove precipitates together with the cell debris. Partition to the interface, frequently observed in experiments with whole cells or large particles like chromosomes [8], appears to be rare if broken cells are employed. In our experiments, cell debris was nearly always suspended in either one of the phases (or both). As can be seen from Table 7, cell debris is normally removed with the lower phase; there is one example (*Brevibacterium ammoniagenes*) where cell debris is removed with the polyethylene glycol-rich upper phase.

In the development of the initial extraction process, several alternative phase systems are often found to be suitable.

The final selection has to take into account several aspects, e.g. the cost of the system, ease of separation (removal of cell debris in an upper phase of higher viscosity is more difficult) and purification factor. (At the clarification of the crude extract often a sizable proportion of other proteins and possible contaminants is removed.) Obviously, the time needed to develop a superior system, once a useful system has been found, also limits the choice. Successful extractions are not necessarily restricted to the conditions given in Table 7. However, this summary demonstrates that

Table 7. Extraction of enzymes from cell homogenates

Organism	Enzyme	Constituent of the phase system	K_{enzyme}	Yield (%)
Candida boidinii	Catalase	PEG 4000/crude dextran	2.95	81
	Formaldehyde dehydrogenase	PEG 4000/crude dextran	10.8	94
	Formate dehydrogenase	PEG 4000/crude dextran	7.0	91
	Formate dehydrogenase	PEG 1000/potassium phosphate	4.9	94
	Isopropanol dehydrogenase	PEG 1000/potassium phosphate	18.8	98
Saccharomyces carlsbergensis	α-glucosidase	PEG 4000/dextran T-500	1.5	75
Saccharomyces cerevisiae	α-glucosidase	PEG 4000/dextran T-500	2.5	86
	Glucose-6-phosphate dehydrogenase	PEG 1000/potassium phosphate	4.1	91
Streptomyces species	Glucose isomerase	PEG 1550/potassium phosphate	3.0	86
Klebsiella pneumoniae	Pullulanase	PEG 4000/dextran T-500	2.96	91
	Phosphorylase	PEG 1550/dextran T-500	1.4	85
Escherichia coli	Isoleucyl tRNA synthetase	PEG 6000/potassium phosphate	3.6	93
	Leucyl tRNA synthetase	PEG 6000/potassium phosphate	0.8	75
	Phenylalanyl tRNA synthetase	PEG 6000/potassium phosphate	1.7	86
	Fumarase	PEG 1550/potassium phosphate	3.2	93
	Aspartase	PEG 1550/potassium phosphate	5.7	96
	Penicillin acylase	PEG 4000/crude dextran	1.7	90
Bacillus sphaericus	Leucine dehydrogenase	PEG 4000/crude dextran	9.5	98
Bacillus species	Glucose dehydrogenase	PEG 4000/crude dextran	3.2	95
Brevibacterium ammoniagenes	Fumarase	PEG 1550/potassium phosphate	0.24	90
Lactobacillus cellobiosus	β-glucosidase	PEG 1550/potassium phosphate	2.2	98
Lactobacillus species	Lactate dehydrogenase	PEG 4000/dextran PL-500[a]	6.3	95

[a] dextran PL-500 was a gift from Pfeiffer & Langen, Dormagen, Germany

a wide variety of enzymes have already been extracted from homogenates of procaryotic and eucaryotic microorganisms, from gram positive as well as from gram negative bacteria. Our earlier observation [39] that the method of cell disintegration had a marginal, if any, influence on the subsequent extraction was reconfirmed for several other cases. Changes in the quality of the biological raw material were also accommodated quite well [10, 14]. The yield remained fairly constant, but the specific activity of the extracted enzyme was influenced as expected. Here, advantage is taken of the fact that the partition coefficient of proteins is widely independent of the total concentration.

4.2 Removal of Interfering Substances

Application of enzymes as industrial catalysts often requires the removal of a small number of other enzymatic activities. The specificity of the reaction, e.g. an enzymatic transformation, is usually determined by the addition of only one substrate. Other enzymes present in the catalyst preparation do not act and may even serve as protecting agents, as long as they do not degrade or alter the product or substrate in an unwanted fashion or digest the catalyst. For example, aspartase present in a fumarase preparation exhibits no interfering activity as long as no ammonium ions are present. In contrast, fumarase has to be carefully removed from an aspartase preparation, since it will react with fumaric acid to yield L-malate by addition of water to the double bond. Since water is necessarily present in the reaction medium, the fumarase level has to be controlled in aspartase preparations. Figure 11 shows that

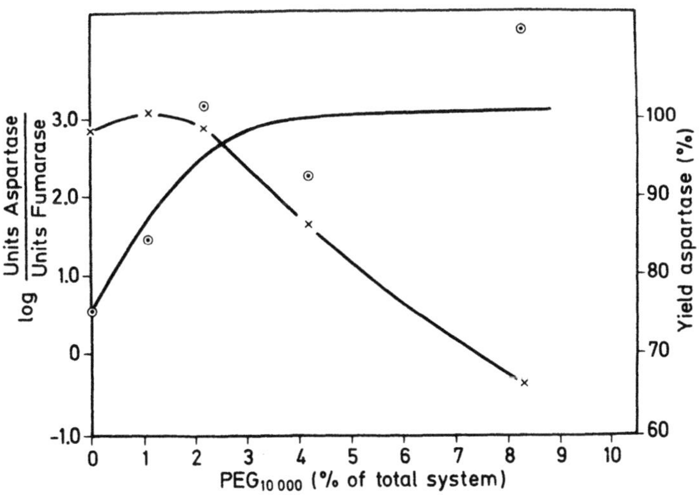

Fig. 11. Separation of aspartase from fumarase by the addition of PEG 10000 to a PEG/phosphate system (system composed of 13.3% potassium phosphate and 50% top phase obtained from the extractive removal of *Escherichia coli* cell debris using a PEG/crude dextran system).
× = logarithm of the activity ratio aspartase/fumarase in the top phase
⊙ = yield of aspartase in the top phase (activity ratio aspartase/fumarase was 0.23 in the top phase of the first extraction)

partition conditions could be found discriminating between aspartase and fumarase from *E. coli* by a factor of 4000. This may reflect exceptionally favourable circumstances, but it demonstrates that consequent development of a phase system can allow very selective separations. In such a case, the final contamination of fumarase in the aspartase preparation will most likely be limited by the degree of separation of the lighter phase obtained at a technical scale. At 99% purity of the aspartase-containing phase, separation drops to 400 fold. To achieve the full potential of partition, separation of liquid phases must also be optimized. This aspect is discussed in Sect. 6.2.

In contrast to the situation discussed above for industrial catalysts, application of enzymes in the analytical or even pharmaceutical field requires a much higher final degree of product purity. Such high degrees of purity cannot be achieved by the simple methods of single stage extractions. Attempts to improve the selectivity of partition have been discussed above but it is also possible to follow conventional chromatographic methods after an initial extraction. In this context, it should be pointed out that a number of undesirable by-products in crude extracts of microorganisms, such as polysaccharides, which lead to viscous solutions, or nucleic acids, which interfere with ion exchange chromatography, can be removed to a large extent by simple partition steps. Since polysaccharides and nucleic acids are more hydrophilic in nature, they stay in the salt phase of a polyethylene glycol/salt system under conditions where proteins are shifted in high yield to the polyethylene glycol-rich phase [10,14] (compare Tables 8 and 9). Also, PEG-dextran systems can be used successfully for the separation of enzymes from nucleic acids and contaminating proteins. Examples from the literature, in which such partition steps are included in enzyme purification procedures, are summarized in Albertsson's monograph [8].

4.3 Examples of Enzyme Purification by Subsequent Partition Steps

We expect that a large number of intracellular enzymes for use as industrial catalysts can be prepared by a series of single step extractions. This would involve only partition steps and liquid-liquid separation, which can be conducted very fast and with high efficiency. Figure 12 illustrates an example of this approach for the isolation of formate dehydrogenase from *Candida boidinii*, presently carried out at the 100–150 g scale of enzyme (calculated as pure protein) in a single experiment, starting from 50 kg of methanol-grown yeast [14]. Table 8 summarizes the analytical data for the purification. The whole procedure is carried out conveniently in less than 4 days. The final catalyst preparation is 70% pure formate dehydrogenase obtained in an overall yield of 70%. The whole process after cell disintegration is carried out at room temperature. There is only one centrifugal separation which considerably reduces the energy demand compared to conventional technology. The formate dehydrogenase prepared has been successfully applied for continuous NADH regeneration in a membrane reactor [40,57]. Major technical problems are not anticipated during further scale-up. We are left with the choice of working in larger batches or developing a continuous process, which also appears quite feasible.

A list of proteins which were partially purified in our laboratory or in the pilot plant by a series of subsequent partition steps is shown in Table 9. The final yield was usually in the range of 70–80% and the purification factor (related to overall protein)

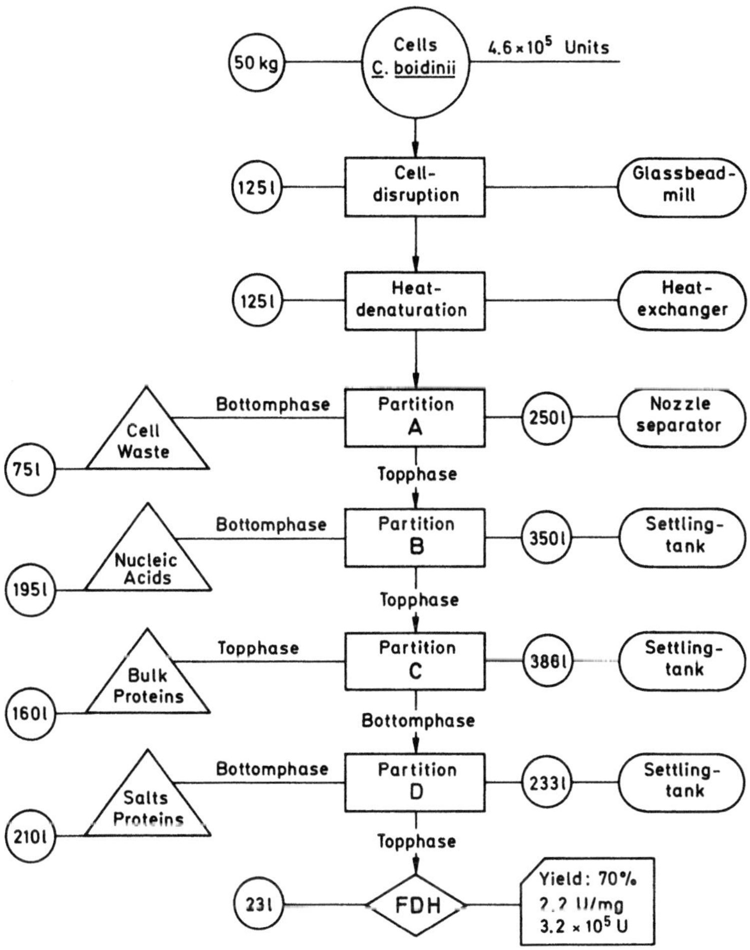

Fig. 12. Flow sheet of the extractive purification of formate dehydrogenase from *Candida boidinii* on a large scale (from Ref. [14])

Table 8. Purification of formate dehydrogenase by liquid-liquid partition (from Ref. [14])

Step	Method	$Yield_{FDH}$ %	Specific activity $U\ mg^{-1}$	Yield of polysacch. %	Yield of "nucleic acid" %	Total activity FDH U
1	Cell disruption	100	0.59	100	100	456 000
2	Heat denaturation	100	0.96	100	100	456 000
3	Top phase A	94	1.08	89	90	429 000
4	Top phase B	75	1.20	31	30	340 000
5	Bottom phase C	70	2.15	11	19	320 000
6	Top phase D	70	2.22	8	4	320 000

50 kg wet cells of *Candida boidinii* were processed

Table 9. Examples of enzymes partially purified by one or several subsequent partition steps in the department "Hochmolekulare Naturstoffe", GBF, D-3300 Braunschweig

Enzyme	Organism	No. of partition steps	Overall purification factor	Overall yield %	Remarks	Ref.
Pullulanase	*Klebsiella pneumoniae*	4	6.3	70	70% pure enzyme, removal of α-amylase and proteases	[10]
α-1,4-glucan phosphorylase	*Klebsiella pneumoniae*	2	2.5	81	removal of glucosyl transferase	[9]
Fumarase	*Brevibacterium ammoniagenes*	3	8.8	70	removal of polysaccharides	[13]
Aspartase	*Escherichia coli*	3	23	68	removal of interfering fumarase	[10]
Glucose dehydrogenase	*Bacillus* species	3	33	83	removal of nucleic acids and polysaccharides	[10]
Leucine dehydrogenase	*Bacillus sphaericus*	2	7.7	97		[10]
Formate dehydrogenase	*Candida boidinii*	4	4.4	70	70% pure enzyme, removal of polysaccharides and nucleic acids	[14]
Glucose isomerase	*Streptomyces* species	1	2.5	86	removal of cell debris under gravity	[10, 5]
β-glucosidase	*Lactobacillus cellobiosus*	1	2.4	98	removal of interfering enzymes	[10]
Interferon	human fibroblasts	1	350	86		[10, 29]

[a] The first partition step always aimed to remove broken cells (with the exception of interferon)

varied between 2.5 and 33. An outstanding result was obtained for the extractive purification of interferon β, where a 350-fold enrichment could be achieved with an activity yield of 86% by a single partition step [29]. Furthermore, it should be pointed out that most of the processes developed for the extraction of the enzymes listed in Table 9 are not fully optimized with regard to the removal of contaminating proteins since, in several cases, decisive parameter was the removal of interfering activities or other unwanted by-products of growth, such as polysaccharides (compare 4.2). Taking this into account, the data presented in Table 9 demonstrate the versatility and the potential of the extraction method for enzyme purification.

4.4 Scale-up Considerations

The scale of extractive enzyme purification has been varied from gram quantities up to 50 kg of wet cells. Increasing the relative proportions, identical results within experimental error are usually obtained provided adequate mixing is maintained. This is no minor achievement, since equilibrium between the phases has to be reachied, which is a diffusion linked process, and diffusion in turn is slowed down by the high viscosity of the phases employed. It is somewhat surprising, therefore, to find that equilibrium is normally reached within a matter of minutes under such circumstances [41]. This is above all due to the unusually low interfacial tension of such systems (for the more viscous dextran-containing systems approximately 10^{-2} to 10^{-3} mN m^{-1}, for polyethylene glycol-salt systems approximately 10^{-1} to 10^{-2} mN m^{-1}) [8]. Therefore, very small droplets are easily generated in the dispersion with minimal energy input during mixing. Beside this, the constance of the partition coefficient while the concentration and process scale vary is an important factor for the scale-up of enzyme purification procedures by partition. Both facts contribute to rater precise calculations of large-scale processes from laboratory data.

This is evident from Table 10, where the yield of formate dehydrogenase obtained in 10 ml scale partition steps is compared with the yield observed in process scale during the enzyme purification discussed in Sect. 4.3. The data are consistent, even at the highest scale-up factor close to 40,000. Analogous results are also shown in Table 13 (see Sect. 6.2).

It should also be mentioned here that scale-up is facilitated further by the availability on the market of the necessary equipment and machinery due to the high standard of extraction technology in the chemical industry, e.g. separators in

Table 10. Performance of scale-up for the partition steps

Step No.	Yield of FDH in 10 ml scale (%)	Yield of FDH in process scale (%)	Scale up factor
A	95	94 (250 l)	25000
B	84	80 (350 l)	35000
C	93	93 (386 l)	38600
D	100	100 (233 l)	23300
Overall yield	74	70	$2\text{–}4 \times 10^4$

a range of different sizes (see Sect. 6.2). Such equipment can be used — sometimes with minor modifications (compare Sect. 6) — when working with aqueous phase systems.

5 Multistage Extractions

5.1 General Comments

In cases where sufficient selective conditions of extraction cannot be found with reasonable effort, multistage operations have to be considered in order to achieve the desired yield or desired purity of the product. If a small number of steps are sufficient, this can be accomplished by repeating the extraction with an appropriate clean phase and separating the phases in the same equipment. A much higher resolving power can be expected if multiple, automatically repeated partitions are performed in a suitable extraction unit. Albertsson described a countercurrent distribution apparatus with 60 transfer steps operating in a Craig unit which can be used with polyethylene glycol/dextran systems [8]. The height of the chambers was minimized to reduce the separation time. Because of the limited volume and capacity, the instrument is mainly used for analytical purposes and separation of cells and cell organelles. For continuous operation in preparative and technical scale, several machines of different designs are available. We tested some of them to measure the extent of performance limitation due to variations in the physical properties of phase systems.

5.2 Mixer Settler

A small mixer settler device [42] has been operated successfully with residence times of about 3 min for extraction with polyethylene glycol/salt systems [43]. Some difficulties were experienced in achieving a tight seal in the all glass apparatus. This may be attributed to the low interfacial tension and lubricating properties of the polyethylene glycol-rich phase. At a technical scale, utilizing other construction materials, the seal should present no difficulty. The density difference is important for the positioning of the overflow, which needs quite precise control in order to maintain a stable, long-term operation.

5.3 Extraction Columns

Several extraction columns differing mainly in the manner by which dispersion and mixing of the phases is accomplished, are described in the literature and are actually used in industry [44,45]. Columns can be operated with countercurrent flow of both phases or by keeping one phase stationary, preferably the heavier phase, while the lighter phase is mobile. When operating an extraction column with aqueous two-phase systems, the problem is not the generation of sufficient exchange surface between phases as in the case of organic solvents, but the avoidance of very small droplet formation which will decrease the performance and lead to flooding of the column.

For experimental purposes, mixing can best be controlled in extraction columns with rotary parts according to the design of Scheibel or Kühni [44]. A schematic drawing of such a column and a flow sheet for operation are given in Figs. 13 and 14. Alternate mixing and settling chambers are employed to allow some coalescence. Blomquist and Albertsson were first to describe the performance of such a column for the separation of nucleic acids, proteins and even particles in two-phase systems composed of PEG 6000 and dextran T-500 [46]. The stirrer speed was reduced to 60 to 80 rpm. A maximum feed rate of 0.12 ml per min and 0.15 ml per min respectively, are reported for countercurrent flow or mobile upper phase under such conditions. While the efficiency of separation was very good, the low feed rate obviously limits the application of PEG/dextran systems in such columns for preparative purposes. A newly developed so called "planet coil centrifuge" may overcome some of these difficulties, at least for operation with a stationary dextran phase [47]. The capacity of extraction columns is considerably improved by utilizing PEG/salt systems, due to the faster phase separation as discussed below.

First results show that 8–10 ml per min could be applied to a column of 2 cm diameter and 200 ml volume. The separation capacity of the column was estimated to reach 250–1000 mg protein per h or 6–24 g per day, when purifying formate dehy-

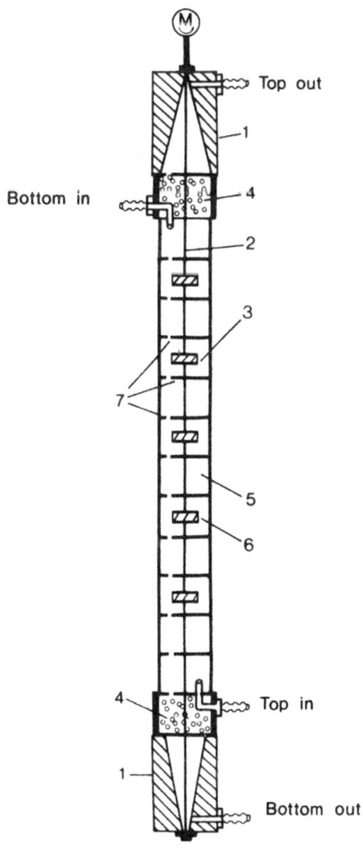

Fig. 13. Schematic diagram of a laboratory extraction column, working with aqueous two-phase systems.
1 outlet chambers;
2 stirrer shaft;
3 stirrer elements (turbine type);
4 settling chambers filled with coalescers (glass beads or others);
5 settling compartment;
6 stirring compartment;
7 stainless steel sieves (free area of about 26%);
M motor drive

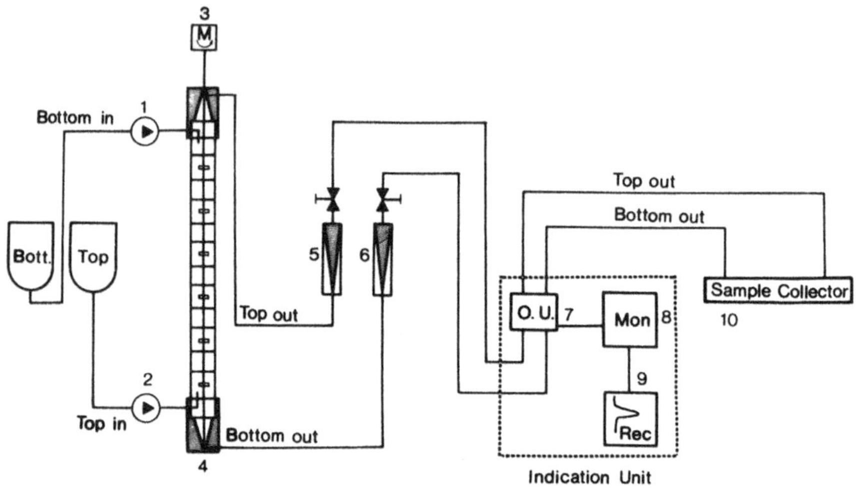

Fig. 14. Flow sheet of the experimental set-up for the laboratory extraction column, working with aqueous phase systems.

1, 2	metering pumps	7	optical unit (photometer)
3	stirrer	8	monitor
4	column	9	recorder
5, 6	flow meter and regulator	10	collector

drogenase or separating catalase and cytochrome C [43,48]. Evaluation of the extraction was performed according to Scheibel [49] applying Eqs. (9) and (10)

$$\frac{C_e}{C_i} = \frac{E - 1}{E^{N+1} - 1} \tag{9}$$

$$N^* = \frac{\log\left[\left(\frac{E-1}{C_f}\right) + 1\right]}{\log E} - 1 \tag{10}$$

where N^* = number of theoretical stages, C_i is the initial solute concentration, C_e the emergent solute concentration, C_f the ratio C_e/C_i, E the extraction factor given by Eq. (11);

$$E = \frac{\text{flow rate of top phase}}{\text{flow rate of bottom phase}} \times K \tag{11}$$

and K is the partition coefficient of the solute. In this way a stage efficiency of 50–60% was attained. The influence of the stirrer speed on the performance of the extraction column is analyzed in Fig. 15. With the particular system at the conditions used, flooding occurs around 200 rpm. The influence of column design, stirrer geometry, sieve plates, etc. has not yet been analysed. Improvements in the operational parameters are also attainable. Since none of the extraction processes using aqueous

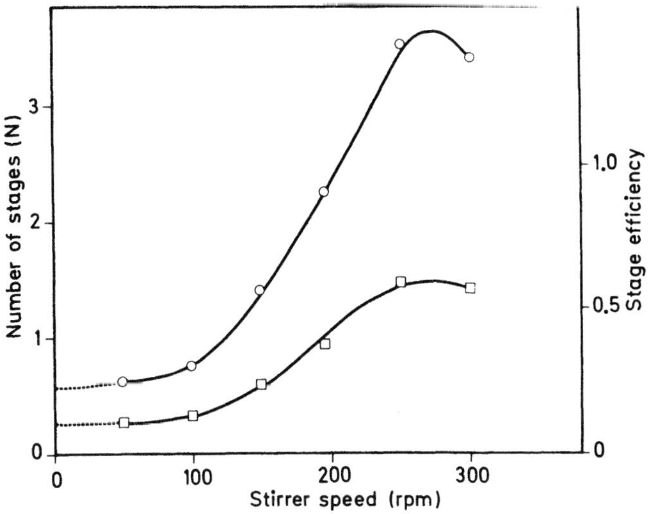

Fig. 15. Dependence of the performance of the laboratory extraction column on the stirrer speed, working with a polyethylene glycol/salt system.
System conditions: 18% w/w PEG 1550, 7% w/w PEG 400, 12% w/w potassium phosphate, pH 7.8
Extraction conditions: test protein was human serum albumin extracted from the salt phase to the PEG-rich phase (K_{HSA} = 1.77); feed ratio — 4 ml per min top phase; 4 ml per min bottom phase. Temperature 21 °C
□ = stage efficiency
○ = number of stages

two-phase systems has been optimized so far, it can be expected that efficiency and capacity could be raised further above the reported levels. The first results are very encouraging and show that continuous multistage processes are possible at the level of capacity for production purposes. Thus, even the final purification of the desired enzyme may eventually be performed by liquid-liquid extraction.

5.4 Graesser Contactor

The Graesser contactor [50], which can be viewed as a horizontal rotary extraction column, has been tested for separation of PEG/salt systems. The intensity of mixing in the commercial machine had to be reduced in order to avoid flooding. Utilizing only the small settling chambers at both ends of the contactor was not effective enough to obtain a sufficient purity of the outflowing phases. Therefore, an additional settling chamber was installed in the process streams outside the contactor which improved the performance significantly. With this minor modification, the extractor was operated continuously for prolonged periods without any problem. Extraction of proteins was simulated using dyes, since the capacity of the extractor was too high to allow preliminary experiments with proteins. A feed rate of 6–10 l h^{-1} was possible, and one can expect a capacity of up to 1 kg·protein per day [43].

5.5 Centrifugal Separators

Centrifugal separators of special design are available which redisperse the phases during the passage of the process streams through the extractor with several stages of extraction built into a single machine. Since sedimentation is accelerated by higher g-forces, such devices should also be useful for polyethylene glycol/dextran phases. This has not yet been tried, but experiments in a Podbielniak centrifugal extractor show that PEG/salt systems can be separated with high feed rates [43]. The temperature increase in the process stream during separation is considerably higher than in disc stack separators because of the higher pressure drop during countercurrent flow operation. This temperature increase has to be taken into account regarding the selection of parameters and evaluation of data, since the equilibrium concentrations of the phases as well as the partition coefficient depend on temperature. If the enzyme is stable enough to permit a temperature rise from 20 to approximately 30 °C, use of the Podbielniak extractor appears an alternative method of achieving higher yields in a single run when partition coefficients are unfavourable.

6 Separation of Aqueous Phase Systems

6.1 Phase Separation under Gravity

The separation of a suspension or dispersion can be described by Stokes Law

$$v_g = \frac{D^2 \Delta\varrho}{18\eta} g \tag{12}$$

where v_g is the settling velocity, D the diameter of the particle or droplet size, $\Delta\varrho$ the density difference, g the acceleration due to gravity and η the dynamic viscosity. The diameter of the droplets in the dispersed phase is changing rapidly. If no additional forces are acting to mix and redisperse the phases, coalescence will dominate and lead to the enlargement of the droplets. Reliable and reproducible methods to measure the rate of coalescence and the average droplet size in aqueous phase systems are not yet available. Dispersions of aqueous phase systems cannot be diluted to obtain a better resolution of a swarm of droplets and observations using thin channels appear to be influenced considerably by experimental conditions. In this case small interfacial tensions lead to redispersion in flow systems. Because of the larger droplet size, higher density difference and lower viscosity, PEG/salt systems separate considerably faster than PEG/dextran systems.

The other physical parameters in two-phase systems can readily be measured. The density difference of the phases may be quite small. Usually it is found in the range of 0.05–0.15 g cm^{-3}. Values >0.1 g cm^{-3} are observed in systems containing crude dextran, high concentrations of broken cells or salts. The viscosity of the dispersion ranges between 3 and 15 mPas [18, 39]. Under conditions frequently employed during initial extraction, a peculiar property of the phase system with regard to the viscosity should be noted. The viscosity difference between the dispersion and the lower

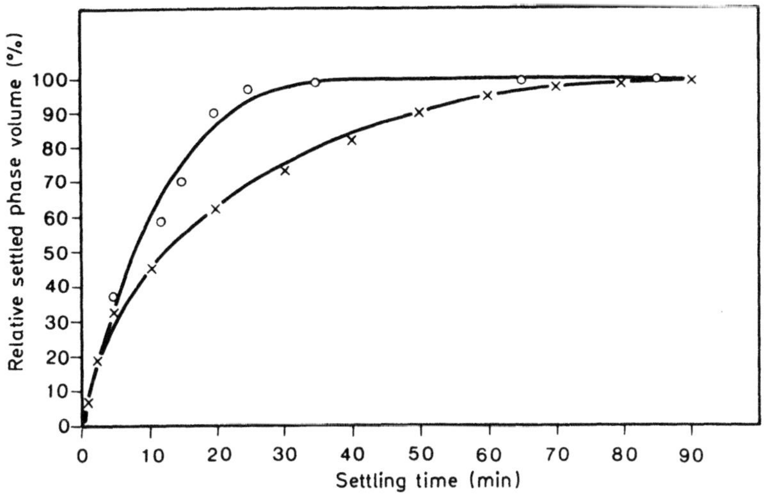

Fig. 16. Settling velocity of two polyethylene glycol/salt systems under gravity used for purification of FDH. Settling tank: glass vessel, 150 l.

× = PEG/potassium phosphate system; 6% PEG 1550, 9% PEG 400, 15% potassium phosphate, pH 7.8 including 2.5 M KCl; top phase dispers, H/D = 2.67; total volume = 150 l (2. step in purification of FDH, Ref. [14]).

o = PEG/potassium phosphate system (12% PEG 4000, 7% potassium phosphate, pH 7.8); bottom phase dispers, H/D = 2.5; total volume = 120 l (2. step in purification of FDH, Ref. [48])

phase can become very large, especially if crude dextran is employed as a phase forming polymer [18] or if PEG/salt systems with high volume ratios are used and heavy precipitates are present. Such systems usually require higher g-forces to speed up separation. In subsequent partition steps, however, liquid-liquid separation can often be performed by gravity in a settling tank if PEG/salt systems are employed [9, 13, 14]. The settling time varies for different systems. Density differences as well as viscosities of the phases (with the exception of salt phases) increase with increasing length of the tie-line. The fastest separation can be expected at intermediate compositions and when the volume of the higher viscosity phase is smaller than that of the lower viscosity phase. Figure 16 illustrates data obtained in settling tanks of similar geometry. Settling times between 30 and 90 min can be sufficient to obtain apparently clear phases [14]. Only careful examination of yield and purity reveals that phases separated under gravity in this way still contain a very small proportion of dispersed phase in minute droplets, which would take much longer periods to completely separate. Separation under gravity can be carried out at any scale without difficulty and could also be performed continuously in a suitable settling tank or in a mixer settler device when multistage procedures are desirable [43]. For systems with dextrans as phase-forming polymer, longer separation times are necessary for complete separation under gravity. From data reported by Albertsson, settling times should be increased approximately fourfold as compared to PEG/salt systems [8]. Improvements of the process by varying the geometric design, for example the ratio of height to diameter, or by increasing the area equivalent of a settling tank, have not been attempted. The advantages of the use of settling tanks are obvious: the specific energy consumption is zero and the process can easily be automated.

6.2 Phase Separation by Centrifugal Separators

If the separation time under gravity becomes too long for practical purposes, commercially available separators can be utilized to speed up separation [39]. All these machines are operated continuously. Figure 17a presents a schematic drawing of a bowl of a liquid-liquid disc stack separator. The dispersion is introduced at the center and is accelerated during radial distribution in the lower part of the bowl. The liquid rises in the channels of the disc stack. Besides the radius, the number and angle of the discs determine the area equivalent of this type of separator, but these parameters also influence the flow pattern in the stack. The flow pattern is quite important when working with aqueous phase systems because the low interfacial tension easily leads to redispersion. After phase separation the lighter liquid is discharged at U and the heavier liquid at L. Discharge can be accomplished by a centripetal pump under pressure or by free flow into a cover. The radial position of U is normally fixed while the radial position of L can be changed within certain limits to accomodate process liquids with different densities. The purity of underflow and overflow as well as the position of the interface in the disc stack are strongly influenced by the correct choice of the discharge position L. Equations (13) and (14) describe the balance conditions, which also hold for aqueous two-phase systems [51].

$$\varrho_u(r_s^2 - r_u^2) = \varrho_l(r_s^2 - r_l^2) \tag{13}$$

$$r_l = \sqrt{\frac{r_s^2(\varrho_l - \varrho_u) + \varrho_u r_u^2}{\varrho_l}} \tag{14}$$

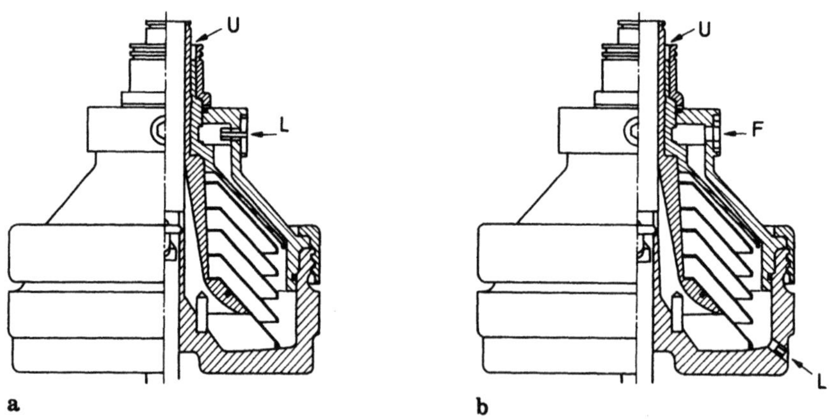

Fig. 17a and b. Bowl of the Gyrotester B-separator;
a for purifier operation (LLO)
U = upper phase outlet
L = lower phase outlet — regulating screw;
b for nozzle operation (NO)
U = upper phase outlet
L = lower phase outlet — nozzle
F = fastener screw

r_u and r_l are the outlet radii of upper phase and lower phase respectively, and r_s the radius of the interface line. ϱ_u and ϱ_l are the densities of upper and lower phases.

Balance conditions have to be met more closely with aqueous phase systems than with organic solvents to achieve optimal results. Therefore, it may sometimes become necessary to make finer adjustments of the radial position of the discharge point of the underflow than anticipated by the manufacturer of the separator. Such changes, however, can easily be accomplished. Since only accessory parts such us gravity discs are involved, no major alteration of the commercial machines are necessary. The delicate balance is also illustrated in Fig. 18, where an optimal flow rate is found with regard to the purity of the discharged phases in an open disc stack separator. The deterioration of the performance at very high feed rates may be expected to be due to the combined effects of several parameters, which limit the throughput according to Eq. (15) [51,52]:

$$\dot{Q} = \frac{D_{lim}^2 \, \Delta\varrho \, g}{18\eta} \, \frac{\omega^2 r}{g} \, F^* \tag{15}$$

where D_{lim} is the limit droplet size to be separated, ω the angular velocity, r the radius of rotation and F^* the effective clarifying surface. The first part of Eq. (15) corresponds to Stokes Law (see Eq. (12)) and the second part to the performance factor of the machine, called the Σ-factor [51,52].

We interpret the increase of dispersed phase in the effluent of an open disc stack separator at lower than optimal feed rates (Fig. 18) as a slight variation in the position of the interface line due to changes in the flow resistance. Such behaviour

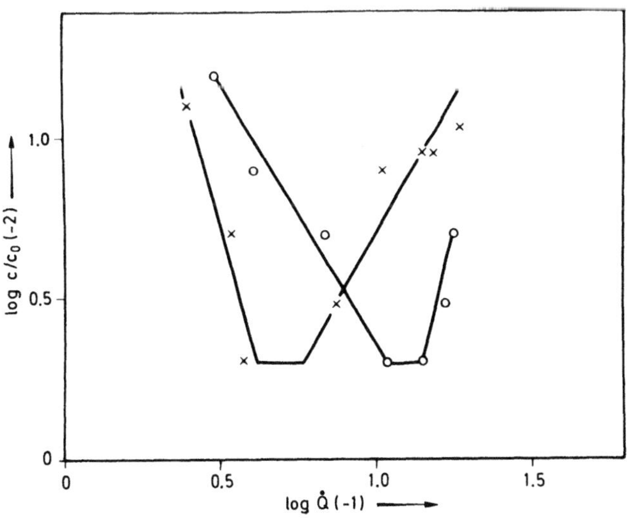

Fig. 18. Throughput characteristics of the α-Laval separator LAPX-202. Logarithmic ratio between the concentration of disperse phase in the effluent (c) and the concentration of disperse phase in the feed (c_0), plotted versus logarithm of the throughput (Q_{lim} — 1 min^{-1}) (from Ref. [48]).
System conditions: 14% PEG 4000, 11% potassium phosphate, 0.1% Blue-dextran as indicator; × at 7000 rpm; o at 9300 rpm

Table 11. Technical data of separators

Separator	Maximum speed	Maximum g-force	Bowl		Disc stack					
			Volume	Inner diameter	Number	Angle	Distance	r_{max}	r_{nin}	Σ
	(rpm)		(1)	(mm)		(degree)	(mm)	(mm)	(mm)	(m^2)
Alfa-Laval[a] Gyrotester B	12000	8500	0.41	106	18	51	1.3	47	21	700
Alfa-Laval[a] LAPX-202	9300	7100	0.35	147	35	51	0.5	51	27	970
Alfa-Laval[a] YEB-1334 nozzle separator	8600	6860	0.82	166	38	51	0.8	56	29	1200
Westfalia[b] SAOH-205	9700	8100	0.25	158	40	53	0.4	52	27	1460

[a] Alfa-Laval — Industrietechnik GmbH., D-2000 Hamburg 80 (FRG)
[b] Westfalia Separator AG., D-4740 Oelde (FRG)

is not observed if discharge takes place under pressure utilizing a centripetal pump [39, 48]. In this case, final adjustment of the back pressure by valves in the discharge line as an additional operating variable leads to conditions where clean phases can be obtained at low feed. Table 11 summarizes technical data for several separators which have been used successfully for the separation of aqueous phase systems. The selection, however, was arbitrary and depended mainly on the availability of equipment of appropriate capacity and size in our pilot plant. So far, we have been able to adapt the operation of any commercial liquid-liquid separator for use with aqueous two-phase systems. Without doubt, separators other than those listed here, can also be utilized.

The interdependence of machine, dispersion and operating parameters is illustrated in Table 12, where the extraction of pullulanase from *Klebsiella pneumoniae* in 3 different phase systems is summarized [18]. Separation of phases were carried out in the Gyrotester operated either as a purifier or with a nozzle bowl (Figure 17b). The separation efficiency can be defined as [18].

$$S_e = P_T - \frac{1 - P_B}{1 + V'_f} \tag{16}$$

where P_T and P_B are the purity factors of overflow and underflow (% pure phase/100), and V'_f the ratio of overflow to underflow $\frac{\dot{Q}_T}{\dot{Q}_B}$. The separation factor can reach a theoretical maximum of 1. A high separation efficiency could be maintained with the purifier bowl while the maximum flow rates were increased due to the influence of $\Delta\varrho/\eta$. Even higher flow rates can be achieved with a nozzle bowl with some loss in the separation efficiency. The separation efficiency in the nozzle separator is clearly related to the viscosity and best results are obtained in the experiment with the highest viscosity of the bottom phase. This means that conditions leading to a high viscosity of the lower phase, e.g. utilization of crude dextran as phase forming polymer, may even be beneficial if fast processing is important [18]. This unexpected result relies on the unusual utilization of a nozzle separator for liquid-liquid separation. This is not a peculiarity of the Gyrotester and Fig. 19 illustrates the performance of another nozzle separator, the YEB 1334. The back pressure at the nozzles must be large enough to keep the interface within the bowl to avoid excessive loss of top phase at the nozzles. The back pressure is influenced by the number and geometry of the nozzles as well as by the viscosity and the rheological properties of the bottom phase. Stable operating conditions are achieved if the ejected heavier phase is equal to the amount of heavier phase delivered in the feed per unit times as defined by Eq. (17) [18]

$$\dot{Q}_n = \dot{Q}_F \frac{1}{V_f} \tag{17}$$

where $\dot{Q}_n$ is the flow through the nozzles, $\dot{Q}_F$ the feed rate to the separator, and V_f the volume ratio of the phases in the feed $\left(V_f = \frac{V_T}{V_B}\right)$.

Table 12. Separation of pullulanase from *Klebsiella pneumoniae* using different types of dextran — in the Gyrotester B-separator, operated with the purifier (LLO) and nozzle (NO) bowl (from Ref. [18])

System	Volumeratio $[V_T/V_B]$	Enzyme yield %	$\dot{Q}_{F\,max}$(LLO) $[l\,h^{-1}]$	Se_{LLO} [a]	$\dot{Q}_{F\,max}$(NO) $[l\,h^{-1}]$	Se_{NO} [a]
9% w/v PEG 4000 2% w/v dextran T-500 (I)	2.50	85	33.0	0.99	87.0	0.88
7 % w/v PEG 400 1.25% w/v crude dextran (II)	2.70	86	37.8	0.99	96.0	0.96
7 % w/v PEG 4000 1.5 % hydrolyzed crude dextran (III)	2.72	86	48.0	0.99	99.0	0.90

All systems containing 0.3 M potassium phosphate, pH 7.8, and 20% wet cells of *Candida boidinii* in a total volume of 20 l;
a Se = separation efficiency factor (see Eq. (16))
System data:

	ϱ_u (g cm^{-3})	ϱ_l (g cm^{-3})	η_u (mPa s)	η_l (mPa s)	$\Delta\varrho$ (g cm^{-3})	$\frac{\Delta\varrho}{\eta}$	Screw length (LLO)	Nozzle (NO)
I.	1.038	1.134	10.0	1100	0.090	0.9×10^{-2}	13.5 mm	2×0.4 mm
II.	1.063	1.159	9.0	3300	0.096	1.1×10^{-2}	13.5 mm	2×0.4 mm
III.	1.076	1.188	8.0	1400	0.112	1.4×10^{-2}	13.5 mm	2×0.4 mm

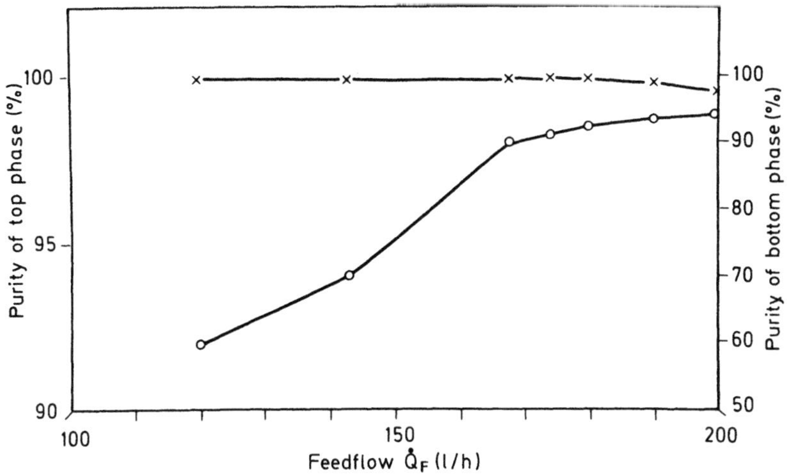

Fig. 19. Performance of separation in the α-Laval nozzle separator YEB 1334, working with a PEG/salt system for separation of formate dehydrogenase from broken cells of *Candida boidinii* (from Ref. [14]). System conditions: 18% PEG 400, 7% PEG 1550, 8% potassium phosphate, pH 7.8, 20% broken cells of *C. boidinii*. × top phase; o bottom phase

If a completely clean top phase is required, the maximum feed rate is limited, which leads to the condition:

$$\dot{Q}_n \leqq \frac{\dot{Q}_{lim}}{V_f} \tag{18}$$

In the nozzle separator YEB 1334 the flow through the nozzles can be manipulated by reducing the number of nozzles from the maximum value of 6 to 4, 3 or 2, or by reducing the diameter of the nozzle orifice. Nozzles with the smallest orifices provided by the manufacturer (0.4–0.5 mm) will usually be sufficient to obtain good results. For the Gyrotester, we had to reduce the orifice from 0.5 mm diameter to 0.4 mm to meet the conditions given by Eq. (18), because with this bowl the number of nozzles (= 2) cannot be changed [18]. If the desired product is in the top phase and cell debris in the bottom phase the nozzle separator can be used with advantage. There is little temperature change during prolonged operation [48]. The top phase was obtained in high purity and good yield. The separation efficiency was quite high and is governed by the purity of the nozzle product. The loss of enzyme containing top phase through the nozzles can be minimized to approximately 1–2%. This mechanical loss is less than the loss of product solution in a packed pellet of cell debris after solid-liquid separation. The nozzle separator has been successfully employed with a bottom phase viscosity >500 mPa s, but the lower limit has not been rigorously tested. The separation becomes much more complicated when cell debris has to be removed with the top phase in an aqueous two-phase system. Such conditions were investigated for the extraction of fumarase from *B. ammoniagenes* [13]. As can be seen from Table 7, fumarase has a partition coefficient of 0.24 and is extracted with the lower phase of a PEG/salt system. In such systems the upper phase, containing the cell waste, exhibits the higher viscosity. During separation, the cell debris should remain suspended in the upper phase and should not collect at the interphase. From the separators tested here for phase separation, the Westfalia SAOH 205 gave the best results. Flow rates of 45–48 $l\,h^{-1}$ could be realized [13]. The separation efficiency for the lower phase was not as high as obtained for extractions into the top phase. For a high yield of fumarase a purity of the lower phase around 95% had to be accepted. This was not detrimental for further processing, however. Residual cell debris was removed in a later extraction step, where separation of phases was carried out under gravity. The possibilities for improving the performance of the separator by variation of machine parameters for such an unfavourable set of dispersion properties are still under investigation.

Performance data for various separators are included in Table 13. The most remarkable fact is the high capacity of liquid-liquid separation. Removal of cell debris and enzyme extraction has been accomplished up to 180 $l\,h^{-1}$ corresponding to a residence time of 16 s. This reflects the dominating influence of the droplet size on the throughput according to Eq. (15). Performance of identical machines for clarification of crude extracts by solid-liquid separation of the same homogenates would be orders of magnitude lower. Therefore, the energy demand for the separation step by liquid-liquid separation is much lower. Not only was the speed of separation outstanding, the desired phase in most instances was obtained in ≧99% purity.

Although extractions and separations described here are performed at room temperature without additional cooling, the activity yields are usually very high. The polymers introduced to establish the two-phase system even appear to stabilize the

Table 13. Performance data of different separators used for liquid-liquid separation

Separator	System	max. Feed rate ($l\,h^{-1}$)	Separation efficiency	τ	Yield of enzyme observed (%)	Yield of enzyme calculated[a] (%)	Enzyme	Ref.
α-Laval Gyrotester B (NO)	PEG-crude dextran	90	0.98	16	96.5	98.6	Formate dehydrogenase	[18]
α-Laval YEB 1334 (NO)	PEG-phosphate	180	0.95	16	94.0	95.0	Formate dehydrogenase	[14]
α-Laval Gyrotester B (LLO)	PEG-crude dextran	48	0.99	30	86.0	88.0	Pullulanase	[18]
Westfalia SAOH-205 (LLO)	PEG-phosphate	45	0.95	20	90.0	93.0	Fumarase	[13]
Westfalia SAOH-205 (LLO)	PEG-phosphate	470	0.99	2	—	—	—	[48]
α-Laval LAPX-202 (LLO)	PEG-phosphate	95	0.99	13	—	—	—	[48]

[a] from 10 ml-experiments; NO = nozzle bowl; LLO = purifier bowl

activity of labile enzymes, so that the customary preventive operation at reduced temperatures can be avoided. This lowers the specific energy demand of extraction processes further. No problems have yet been experienced in breaking a dispersion of aqueous two-phase systems. Even dispersions containing detergent could be handled without difficulties [9,10]. Existing knowledge in chemical engineering can be utilized to analyze the separation of aqueous phase systems and guide the selection of appropriate machine and operational parameters. Commercially available equipment can be employed with no or only minor modifications in accessory parts to achieve optimal results in separation. Further scale-up appears possible without difficulties.

7 Removal of Polymers

If extraction of enzymes is carried out in aqueous two-phase systems, the question arises, how can polymers be removed from the desired product? In most practical cases, this will mean the separation of the enzyme from polyethylene glycol. If the enzyme is extracted into a polyethylene glycol-rich top phase, addition of salt will lead to the establishment of a new phase system. Using appropriate conditions the enzyme can be reextracted and even concentrated into the salt phase [8,9,13,14]. Such procedures would, in principle, also permit the recovery of most of the polyethylene glycol. Residual amounts in the salt phase can either be tolerated or removed by common unit operations such as ultrafiltration or dialysis. Since the average molecular weight of polyethylene glycol used for extraction is low (6000 dalton or smaller), separation by these methods of proteins with $M_r > 60{,}000$ is easily accomplished.

Diafiltration for the separation of proteins can also be carried out directly from PEG-rich phases after sufficient dilution. If the enzyme has a high molecular weight, ultrafiltration membranes with large pore sizes can be utilized, which in turn contribute to a higher flux rate and a further gain in purification by removing smaller proteins. In principle, it should also be possible to separate proteins with $M_r < 300{,}000$ from dextran T-500 by ultrafiltration and recover both, but experience is lacking so far.

Precipitation of the desired enzyme by addition of solvents or salts is possible, but removal of solids is sometimes impaired if the surface properties of the precipitate are changed by the presence of minute amounts of polymers.

Adsorption of the enzyme to ion exchangers or other suitable adsorbents is simple after adjustment of pH and ionic strength. The phase forming polymers can be washed away and the product later eluted by changing the elution buffer. All three methods are applicable on a technical scale. Ultrafiltration appears to be the method of choice if applicable. It can also be integrated into continuous processes without difficulty.

8 Economic Considerations

A detailed system analysis is not available at present for comparing alternative processes during recovery and purification of intracellular enzymes and the consequences on the cost for the same product quality. As a first, rough approximation we calculated the running cost for three different processes for the purification of formate dehydrogenase from *Candida boidinii* as given in Fig. 20 [14]. The final yield and purity are presented in Table 14 together with the cost index and the productivity. The substantial savings of the 4-step liquid-liquid separation process compared to conventional techniques arise mainly from the reduction in process time. This directly affects labour costs and energy demand. The expenditures for cooling are not even included in the values given in Table 14. The liquid-liquid process was carried out at room temperature in contrast to the conventional and mixed processes which are conducted in part at 4 °C. Higher activity yield and the reduced process time more than compensate for the extra expenses for the polymers and salts needed to establish two-phase systems.

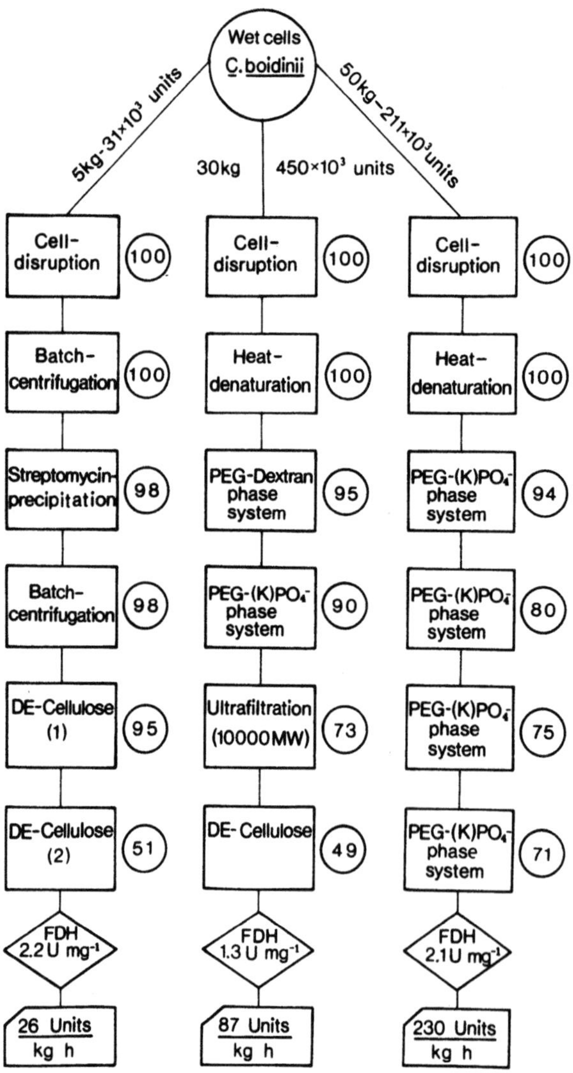

Fig. 20. Alternative process lines for formate dehydrogenase isolation from *Candida boidinii* (from Ref. [14]).
Left column = standard procedure described by Schütte et al. [55]; middle column = mixed procedure; right column = extraction procedure

Table 14. Comparison of different formate dehydrogenase preparations (from Ref. [14])

Method	Total cellmass kg	Initial units	Purity $U\ mg^{-1}$	Yield %	Net time h	Performance factor $U\ kg^{-1}\ h^{-1}$	Cost index[b] DM per Unit
Partition (1)	50	211×10^3	2.1	71	13	230	13×10^{-3}
Partition (2)	50	460×10^3	2.2	70	18	356	7×10^{-3}
Mixed	30	450×10^3	1.3	49	61	120	18×10^{-3}
Standard[a]	5	31×10^3	2.2	51	121	26	374×10^{-3}

[a] Schütte et al. [55]
[b] rough calculation, based on labour, material and running costs (without cultivation of *C. boidinii*)

Actual process costs are strongly influenced by the kind and concentration of the hydrophilic polymers selected. For the reasons discussed above, we have only used PEG and dextran up to now. PEG is available in bulk quantities at reasonable cost. Purified dextran fractions, however, are quite expensive. The prices for individual components of commonly used systems are listed in Table 15. The proportional cost for dextran fractions T-500 are very high and may reach 90% of the extraction medium. The cost can be reduced by substitution of highly purified fractions with crude dextran or hydrolyzed crude dextran [18]. No serious difficulties were encountered while processing phases containing crude dextran. For the first extraction from cell homogenates advantage can be taken of the fact that cell

Table 15. Cost index[a] of PEG 4000-dextran systems, containing 0.3 M potassium phosphate, pH 7.8 (from Ref. [18])

System	PEG 4000		Dextran		Potassium phosphate		Total price	Dextran price / Total price
	$[g\ l^{-1}]$	[DM]	$[g\ l^{-1}]$	DM	$[g\ l^{-1}]$	[DM]	$[DM\ l^{-1}]$	
9 % w/v PEG 4000 2 % w/v dextran T-500 (I)	90	0.45	20.0	6.60	50	0.50	7.55	0.87
7 % w/v PEG 4000 1.25 % w/v crude dextran (II)	70	0.35	12.5	0.88	50	0.50	1.73	0.51
7 % w/v PEG 4000 1.5 % w/v crude dextran hydrolyzed (III)	70	0.35	15.0	1.05	50	0.50	1.90	0.55
9 % w/v PEG 4000 0.5 % w/v crude dextran (IV)	90	0.45	5.0	0.35	50	0.50	1.30	0.27

[a] calculations are based upon bulk prices per 100 kg, in autumn 1980 in Deutschmarks (DM)

debris and other soluble cell constituents are polymers temselves and will influence the carrier system, lowering the critical concentration where phase separation takes place. Under these conditions, very low concentrations, e.g. 0.3–0.5% dextran may be sufficient to generate two liquid phases. This lowers the proportional cost to ~30% when considering single extractions. While the initial supply of PEG can be used for several subsequent extraction steps and may eventually be recovered, crude dextran phases with suspended cell debris have to be discarded. Also, considerable effort was put into the development of alternative PEG/salt systems for enzyme extraction which are not burdened by the cost of purified dextrans. This was successful in several instances (see Table 7). An analysis of the clarification of cell homogenates in order to obtain the so-called "crude extract" showed that the better performance of liquid-liquid separation compared to conventional solid-liquid separation by centrifugation becomes very much apparent with increasing scale of operation which compensates for the cost of the components constituting the phase system [18].

These results demonstrate that the price of the materials is not a priori prohibitive for successful implementation of the technology. The higher activity yield and the higher specific activity of enzyme solutions usually found in two-phase systems are additional advantages which improve the performance of subsequent operations. These effects are not considered in the calculations presented in Table 14. The much shorter residence times in the separation of two-phase systems allow the use of much smaller machines, leading to substantial savings in initial investment and improve the availability of equipment in a multi-purpose plant. It is possible to switch from liquid-liquid extraction to conventional methods in the final purification steps without difficulties, so that the specific advantages of the various technologies can be combined to achieve the best result and most economic process.

The application of aqueous two-phase systems for the extractive purification of enzymes on a production scale can lead to higher waste water loads which must also be taken into account. In our view, this should present only minor problems: the polymers are nontoxic and biodegradable [53]. If the removal of salt becomes necessary using PEG/salt systems, sodium phosphate can be employed, which can easily be removed (and then recovered) from the phase by cooling the solution to 6 °C, which leads to its precipitation [10]. Furthermore, the PEG introduced for the removal of cell debris can be used up to the last partition step with adequate process planning (and can, in principle, be recovered, see Sect. 7). The phase volumes can also be chosen in order to obtain high concentrations, leading to relatively small volumes of the phases to be discarded. Therefore, the problem of waste water should not be an obstacle for the application of the extraction technology.

9 Concluding Remarks

The extraction of enzymes and other biologically active proteins from complex mixtures is an emergent technique especially suited to large scale isolation processes. In this way, difficulties inherent to solid-liquid separation related to small particle sizes can be avoided. Liquid-liquid separation is carried out using conventional equipment and much knowledge available in other fields of chemical engineering can be applied. Substantial savings in process time, energy requirements and labour costs more than compensate for the extra expenses involved in establishing a suitable aqueous two-phase system. Scale-up appears to be straight forward and relatively easy. It has been carried out in our pilot plant with up to several million units or 150 g enzyme protein per charge. The entire process is conducted at room temperature and high yields are usually obtained.

There are several options — not yet sufficiently studied — for the future developments of the basic concept: improvements in the selectivity of extraction, continuous processing and multistage operation. Following such extractions, conventional technology e.g. ultrafiltration or chromatographic separations, can be applied for further processing without difficulties. It appears highly probable that, with increasing demand for intracellular enzymes and other proteins produced in microorganisms by recombinant DNA technology, extraction will become a major and familiar unit operation in the biochemical industry.

10 Acknowledgements

The authors are very grateful to Dr. G. Johansson, Dr. U. Menge, Mr. H. Schütte, Mr. A. F. Bückmann, and Dr. M. Morr for their collaboration and helpful discussions. We are indebted to the technical staff of the department Hochmolekulare Naturstoffe for many contributions to the work described, and to Mrs. H. Rabe for her careful and patient typing of the manuscript.

11 Symbols

C_T	concentration in the top phase ($U\ l^{-1}$, $kg\ l^{-1}$, $mol\ l^{-1}$)
C_B	concentration in the bottom phase ($U\ l^{-1}$, $kg\ l^{-1}$, $mol\ l^{-1}$)
C_e	emergent concentration ($U\ l^{-1}$, $kg\ l^{-1}$, $mol\ l^{-1}$)
C_i	initial concentration ($U\ l^{-1}$, $kg\ l^{-1}$, $mol\ l^{-1}$)
D	diameter of droplet or particle (m)
D_{lim}	limit droplet, (particle) diameter (m)
E	extraction factor
F	Faraday constant (96,485 $C\ mol^{-1}$)
F*	clarifying surface (m^2)
g	earth acceleration ($m\ s^{-1}$)
G	partition ratio
k	Boltzmann constant ($1.381 \times 10^{-23}\ JK^{-1}$)
K	partition coefficient
K_i	inhibition constant ($mol\ l^{-1}$)
K_m	Michaelis Menten constant ($mol\ l^{-1}$)
K_p	partition coefficient of protein
K_p^0	partition coefficient of protein when phase potential or charge of protein is zero
K^-	partition coefficient of small anion
K^+	partition coefficient of small cation
K_E	partition coefficient of enzyme

K_F	partition coefficient of polyethylene glycol dye-derivative
M	molecular weight (g mol^{-1})
$\bar{M}_w$	weight average molecular weight (g mol^{-1})
$\bar{M}_n$	number average molecular weight (g mol^{-1})
M_r	relative molecular weight
N	Newton (kg m s^{-2})
mN	milli Newton (10^{-3} kg m s^{-2})
N^*	number of theoretical stages
n	number of binding sites
mPa	milli Pascal (10^{-3} N m^{-2})
P_T	purity of top phase (%)
P_B	purity of bottom phase (%)
$\dot{Q}_B$	volumetric flow of bottom phase (l h^{-1})
$\dot{Q}_F$	volumetric flow of feed (l h^{-1})
$\dot{Q}_{lim}$	limiting flow capacity (l h^{-1})
$\dot{Q}_n$	volumetric flow through the nozzles (l h^{-1})
$\dot{Q}_T$	volumetric flow of top phase (l h^{-1})
r_s	radius of interphase position (m)
r_u	radius of upper phase outlet (m)
r_l	radius of lower phase outlet (m)
R	general gas constant (8.314 J mol^{-1} K^{-1})
S_e	separation efficiency
T	absolute temperature (K)
U	unit of enzyme activity (μMol min^{-1})
V_T	volume of top phase (l)
V_B	volume of bottom phase (l)
V_f'	volume ratio of overflow and underflow
V_f	volume ratio of upper and lower phase
V_g	sedimentation velocity (m s^{-1})
Y_T	yield of enzyme in top phase (%)
Y_B	yield of enzyme in the bottom phase (%)
Z_p	number of charges of protein
Z^+	number of charges of cation
Z^-	number of charges of anion
α	angle (degree)
ϱ_u	density of upper phase (kg m^{-3})
ϱ_l	density of lighter phase (kg m^{-3})
$\Delta\varrho$	density difference (kg m^{-3})
λ	proportionality factor of interacting forces (J g mol^{-1})
Σ	Sigma factor, for comparism of centrifuges (m^2)
ψ	interphase potential (mV)
ω	angular velocity (rad s^{-1})
η	dynamic viscosity (kg s^{-1} m^{-1})
$[\eta]$	limiting viscosity (kg s^{-1} m^{-1})
τ	residence time (s)

12 References

1. Loncin, M.: Die Grundlagen der Verfahrenstechnik in der Lebensmittelindustrie, Sauerländer, Aarau u. Frankfurt/M. 1969
2. Higgins, J. J. et al.: Biotech. Bioeng. *20*, 159 (1978)
3. Mosqueira, F. G. et al.: Biotech. Bioeng. *23*, 335 (1981)
4. Kula, M.-R. et al.: German Pat. 2639129, US-Pat. 4,144,130
5. Kula, M.-R.: Extraction and Purification of Enzymes in: Applied Biochem. and Bioeng. (Wingard, L. B. jr., Katchalski-Katzir, E., Goldstein, L., eds.), Vol. 2, New York: Academic Press, New York 1979, p. 71
6. Beijerinck, M. W.: Zentr. Bl.-Bakt. *2*, 627, 698 (1896)
7. Beijerinck, M. W.: Kolloid-Z. *7*, 16 (1910)
8. Albertsson, P. Å.: Partition of Cell Particles and Macromolecules, 2nd ed., Wiley Interscience, New York 1971
9. Hustedt, H. et al.: Biotech. Bioeng. *20*, 1989 (1978)
10. Hustedt, H. et al.: unpublished results
11. Kula, M.-R., Johansson, G., Bückmann, A. F.: Biochem. Soc. Trans. *7*, 1 (1979)
12. Kroner, K. H. et al.: in: Affinity Chromatogr. and Related Techniques, (Gribnau, T. C. J., Visser, J., Nivard, R. J. F., eds.), p. 491. Elsevier, Amsterdam 1982
13. Hustedt, H., Kroner, K. H., Kula, M.-R.: Large-Scale Isolation of Fumarase by Partition, Poster presented at the 1. Eng. Foundation Conf. on Advances in Fermentation Products Recovery Process Techn., Banff, Canada 1981
14. Kroner, K. H. et al.: J. Chem. Techn. Biotechn., *32*, 130 (1982)
15. Vollmert, B.: Polymer Chemistry, Springer, Berlin 1973
16. Brønstedt, J. N. Z.: Z. phys. Chem. A., 2571 (1931) (Bodenstein-Festb.)
17. Brønstedt, J. N. Z., Warming, E.: Z. phys. Chem. A *155*, 343 (1931)
18. Kroner, K. H., Hustedt, H., Kula, M.-R.: Biotechn. Bioeng., *24*, 1015 (1982)
19. Hustedt, H., Kula, M.-R.: Eur. J. Biochem. *74*, 191 (1977)
20. Foster, P. R., Dunnill, P., Lilly, M. D.: Biochim. Biophys. Acta *317*, 505 (1973)
21. Hönig, W., Kula, M.-R.: Analyt. Biochem. *72*, 502 (1976)
22. Johansson, G.: Biochim. Biophys. Acta *221*, 387 (1970)
23. Ricketts, C. R.: Progr. Org. Chem. *5*, 73 (1961)
24. Albertsson, P. Å.: Endeavour *1*, 69 (1977)
25. Schönfeldt, N.: Grenzflächenaktive Äthylenoxid-Addukte, Wissenschaftl. Verlagsges., Stuttgart 1976
26. Bückmann, A. F.: German Pat. 2841414
27. Bückmann, A. F., Morr, M., Johansson, G.: Makromol. Chem. *182*, 1379 (1981)
28. Morr, M.: unpublished results
29. Menge, U. et al.: German Pat. Application, Offenlegungsschrift DE 2943016A1
30. Flanagan, S. D., Barondes, S. H.: J. Biol. Chem. *250*, 148 (1975)
31. Hubert, P. et al.: FEBS Letters *65*, 169 (1976)
32. Takerkart, G., Segard, E., Monsigny, M.: FEBS Letters *42*, 218 (1974)
33. Mosbach, K.: Biochem. Soc. Trans. *2*, 1294 (1974)
34. Ohlsson, R., Brodelius, P., Mosbach, K.: FEBS Letters *25*, 234 (1972)
35. Easterday, R., Easterday, I. Affinity Chromatogr. of Kinases and Dehydrogenases on Sephadex and Sepharose Dye Derivatives in: Immobilized Biochemicals and Affinity Chromatogr., (Dunlop, R. ed.), p. 123. Plenum, New York 1974
36. Amicon Corporation "Dye-Ligand Chromatogr.", Lexington/USA 1980
37. Kula, M.-R. et al.: FEBS Special Meet. on Enzymes, Abstract S 6–30, Dubrovnik/Yugoslavia 1979
38. Johansson, G., Hartmann, A., Albertsson, P. Å.: Eur. J. Biochem. *33*, 379 (1973)
39. Kroner, K. H. et al.: Biotech. Bioeng. *20*, 1967 (1978)
40. Leuchtenberger, W. et al.: 2nd Europ. Congr. of Biotechn., (Abstracts of Communications No 47), Eastbourne 1981
41. Shanbag, V. P.: Biochim. Biophys. Acta *320*, 517 (1973)
42. Müller, E., Kreichelt, H. J.: Vt-Verfahrenstechnik *13*, 142 (1979)

43. Hustedt, H. et al.: in: Enzyme Eng. (Weetall, H. H., Royer, G. P., eds.), Vol. 5, p. 45. Plenum, New York 1980
44. in Treybal, R.: Liquid Extraction, 2nd ed., Mc. Graw Hill, New York 1963
45. Hanson, C.: Recent Advances in Liquid-Liquid Extraction, Pergamon, Oxford 1971
46. Blomquist, G., Albertsson, P. Å.: J. Chromat. *73*, 125 (1972)
47. Ito, Y. et al.: Sep. Sci. Technol. *15*, 1589 (1980)
48. Kula, M.-R. et al.: in: Biochem. Eng. II (Constantinidis, A., Vieth, W. R., Venkatasubramanian, K., eds) Annals N.Y. Academy of Science, Vol. *369*, 341 (1981)
49. Scheibel, E. G.: in: Techniques of Organic Chem., 2nd ed., (Weißberger, A. ed.), Vol. *3*, part 1, p. 332. Interscience, New York 1956
50. Coleby, J.: Brit. Pat. 860880, 972035, 1037573
51. Hemfort, H.: Verfahrenstechnik *4* (5), 167 (1970)
52. Murkes, J.: Filtration Separation, Mar./Apr., 112 (1966)
53. Cox, D. P.: The Biodegradation of Polyethylene Glycols in: Advances in Applied Microbiology (Perlman, D., ed.), Vol. 23, p. 173. Academic Press, New York 1978
54. Bückmann, A. F., Morr, M., Kula, M.-R.: IUPAC 26th Int. Symp. on Macromolecules, Short communications, (Lüdenwald, I., Weis, R., eds.), Vol. III, p. 1551, 1979
55. Schütte, H. et al.: Eur. J. Biochem. *62*, 151 (1976)
56. Hustedt, H. et al.: First Europ. Congr. on Biotechn. Interlaken, Preprints, Part 1, p. 48, 1978
57. Wichmann, R. et al.: Biotech. Bioeng. *23*, 2789 (1981)

Biomass Separation from Liquids by Sedimentation and Centrifugation

Udo Wiesmann, Herbert Binder
Institut für Chemieingenieurtechnik, Technische Universität Berlin,
Ernst-Reuter-Platz 3–5, 1000 Berlin 10, FRG

Sedimentation tanks and centrifuges are used in several biotechnological processes to separate the produced biomass from culture media. After reviewing different types of separators, methods of calculation are described, both for separation in low concentration regions (clarification) and in higher ones (thickening). Firstly, closed systems are considered which are useful for the measurement of settling rates. The hydrodynamic theory for free settling of single particles as well as the theory of the limiting flux for higher particle concentrations are discussed. Differences between ideal suspensions of rigid particles and non-ideal suspensions of flocks which were formed in many bio-suspensions are pointed out. Methods of calculation for continuous clarification and thickening in vertical flow tanks and tube centrifuges with continuous solid discharge (decanter) are described. As far as possible theoretical results are compared with experimental ones. Some examples for the application of sedimentation tanks and centrifuges in biotechnology are given. Finally some remarks about flocculation processes are added.

1 Introduction

In most bioreactors microorganisms are suspended in liquid culture media as single cells, flocks or particles formed by mycelia. After completing the reaction they must be separated by solid-liquid separation processes such as sedimentation, centrifugation, flotation or filtration. In continuous processes, sedimentation and centrifugation are preferred as first separation steps. If possible they should be carried out continuously. Part of the thickened biomass is often recycled to the bioreactor to obtain high efficiencies. Examples for technical applications are:

— the activated sludge process for waste water treatment,
— the anaerobic digestion of waste water and
— the continuous production of single cell protein (SCP).

A further application is the continuous production of ethanol as a liquid fuel, but until now this type of process has not reached commercial scale. The excess thickened biomass can be separated from the liquid in further steps by flotation and/or surface filtration. This also will be often necessary if the excess biomass is only an undesired by-product as in biological waste water treatment processes.

Each suspension is separated by sedimentation or centrifugation in two flows with lower solid concentration $V_K c_K$ and a higher one of $\dot{V}_D c'_D$ (Fig. 1). In liquids containing lower solid concentrations the separation process is called *clarification*. Compared to it the process is called *thickening* if regions with higher solid concentrations are present.

Fig. 1. Clarification and thickening — two parts of solid/liquid separation processes —

In technical solid-liquid separation processes three conditions can exist:

1) Liquid as a primary product. The solid concentration of the overflow is then to be as low as possible. Since clarification is the most important process here the apparatus is often called a *clarifier*. The clarification ratio

$$\beta = \frac{\dot{V}_0 c_0 - \dot{V}_K c_K}{\dot{V}_0 c_0} \approx 1 - \frac{c_K}{c_0} \tag{1}$$

has to be as near as possible to one. In this case sludge is the secondary product. Its high water content causes additional costs for transport and deposition.

2) The solid contains the primary product and its concentration attained should be as high as possible. Thickening is then the most important process and the apparatus is often called a *thickener*. The thickening ratio

$$n_E = \frac{c'_D}{c_0} \quad (2)$$

should be as high as possible. Here water is the undesired secondary product. It is often discharged in rivers or lakes without being sufficiently free of suspended solids.

3) In order to avoid environmental damages the simultaneous treatment of secondary products is of increasing importance. In this case, both a high clarification ratio and a high thickening ratio are desired. Therefore, the mechanism of particle settling in the low as well as in the high concentration range must be considered in calculation and operation methods to attain correct dimensioning of the apparatus and optimal process control.

2 Sedimentation

2.1 Sedimentation Tanks

The principle of operation consists of utilizing solid-liquid separation from the relative motions of solid particles and liquid due to gravity. Sedimentation processes can be classified due to horizontal, vertical or inclined direction of liquid flow. Three types of conventional tanks are mainly used: the rectangular horizontal flow tank, the circular radial flow tank and the circular or square upward vertical flow tank (frequently in a hopper form). Furthermore, some inclined lamella separators have recently achieved technical application.

2.1.1 Rectangular Horizontal Flow Tanks

The suspension flows in a longitudinal direction. Homogeneous flow is approximated by using dash-plates or perforations installed in downflow baffles (Fig. 2a). At the exit of the tank the overflow flows over a weir. By using this kind of discharge, secondary circulation flow patterns cannot be avoided. Smaller tanks are equipped with several sludge hoppers for the thickening and storage of the sludge (Fig. 2b). From the lowest points of the hoppers the underflow is drained off by pumps or by hydrostatic sludge removal. For flow rates above 50 $m^3 h^{-1}$ which until now have only been found in waste water treatment processes, only one sludge hopper is installed near the tank inlet. The bottom of the tank is mechanically cleared by sludge scrapers. Two different systems are used: a continuous chain and flight collecting mechanism (Fig. 2c) and a discontinuous mechanism with discharge cars and rake arms (Fig. 2d).

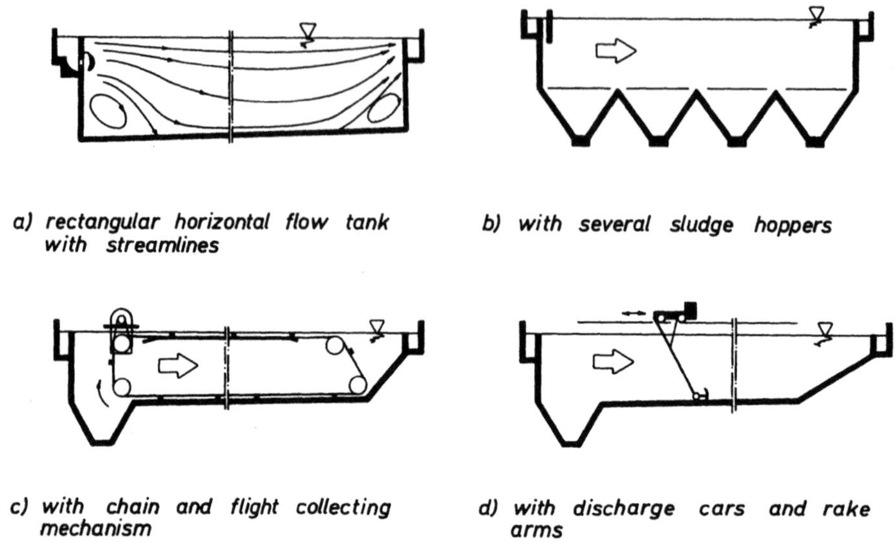

Fig. 2a—d. Rectangular horizontal flow tanks

2.1.2 Circular Horizontal Flow Tanks

A circular horizontal flow tank is flat in relation to its diameter and is used for high flow rates. The suspension is fed into the center of the tank, and the overflow is drawn off at the maximum radius by an overflow weir. Since the angle of the slope of the bottom is small, rotating sludge scrapers are necessary. They convey the settled sludge helically to a sludge hopper at the center of the tank. Two different kinds of drive are used: a central drive and bearing with scrapers in a wing-truss structure (Fig. 3a) and a peripheral drive and bearing with a raking sear bridge (Fig. 3b).

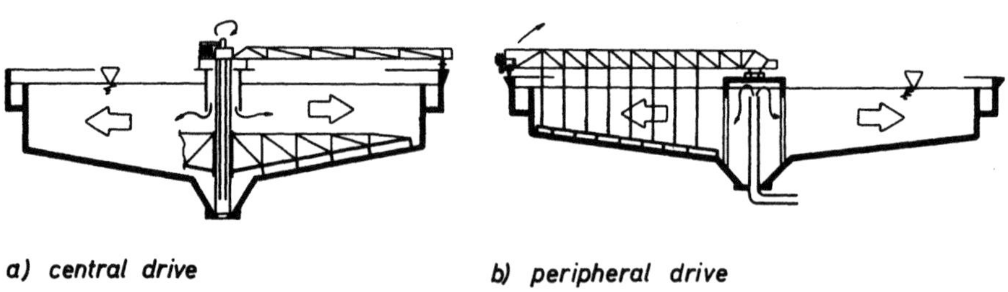

Fig. 3a and b. Circular horizontal flow tanks

2.1.3 Circular or Square Vertical Flow Tanks

Vertical flow tanks are high in relation to their diameters and are used for smaller flow rates. The suspension is fed into the center of the tank, and the overflow is often drawn off by an overflow weir at the maximum radius. However, several spillway channels are sometimes installed in the upper cross section to cause an approximate homogeneous vertical flow. Several different types are constructed: a cylindrical form with a slightly inclined bottom and a rotating sludge scraper (Fig. 4a), a tank with an upper cylindrical and lower conical part (Fig. 4b), and a conical tank called the "Dortmund tank" (Fig. 4c). The last two types are built at the present time with an upper square cross section and a pyramidal hopper. Particulary in connection with neighbouring bioreactors other different special types are often used (Fig. 4d). For favourable operation conditions and suitable properties of the suspended particles, a sludge blanket forms in the upper part of the tank. Smaller particles which would otherwise escape are removed by fluidized bed filtration or flocculation.

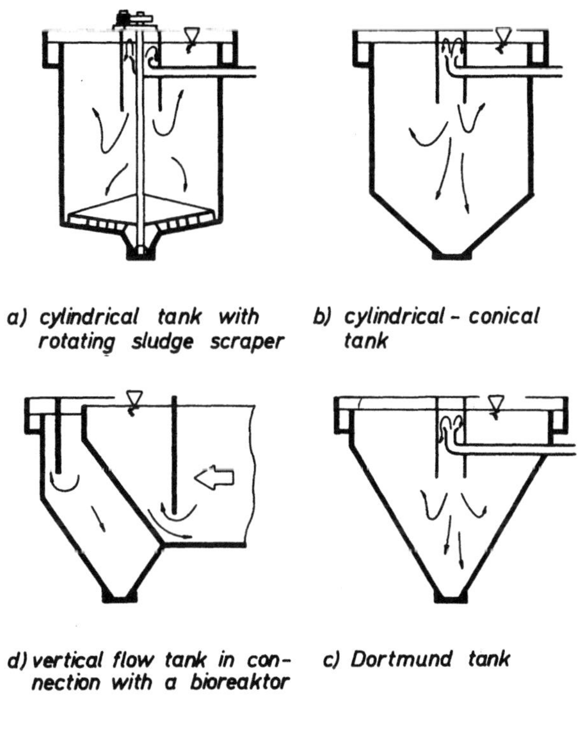

Fig. 4a—d. Vertical flow tanks

2.1.4 Lamella Type of Separators

Lamella type of separators are packings of parallel lamellas of different shapes. Usually plates, tubes and structures with honeycombed and wavelike cross sections made of plastics are used (Fig. 5a). According to the directions of the liquid flow and the flow of settled solids, three different methods of operation can be distinguished: direct current, countercurrent and crosscurrent (Fig. 5b). Countercurrent separators with parallel plates have recently been successfully used. In all practical solutions the avoidance of resuspension of settled solids is of fundamental importance. Therefore, the suspension is frequently fed sideways into the lower part so that an intermediate type of counter- and crosscurrent operation results (Fig. 5c).

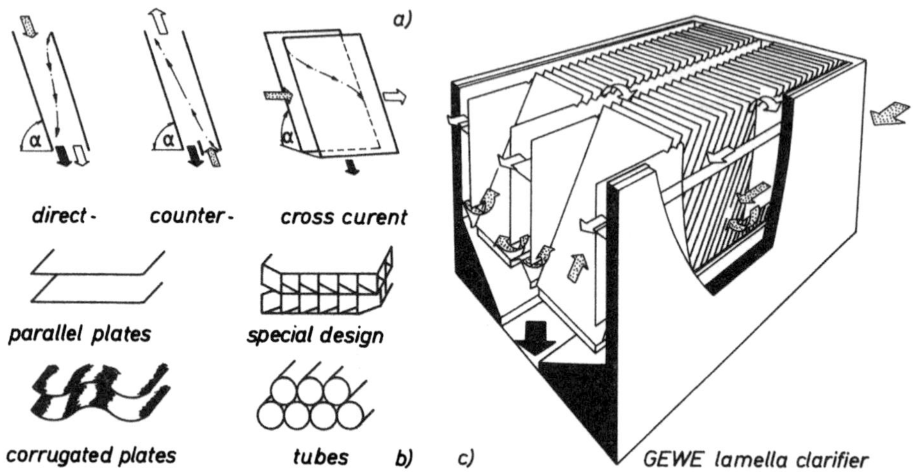

Fig. 5a—c. Lamella separators, Fig. 5c reprinted by permission of GEWE Water Engineering, Lund, Sweden

2.2 Sedimentation in Closed Systems

2.2.1 Sedimentation in the Range of Low Particle Concentration

In the range of low particle concentration, sedimentation can be described as the sedimentation of single particles which do not influence the settling rate of neighbouring particles. For spherical particles, the settling rate w_{s0} can be calculated on the basis of hydrodynamic methods.

The following forces influence the settling rate:

$$\text{gravitational forces} \qquad F_g = \varrho_p V_p g \tag{3}$$

$$\text{buoyant forces} \qquad F_b = \varrho V_p g \tag{4}$$

and

$$\text{drag forces} \qquad F_d = \frac{\zeta \varrho w_{s0}^2}{2A_p} \tag{5}$$

where ϱ_p = particle density
ϱ = liquid density
V_p = particle volume
A_p = cross section area of the particle
g = acceleration due to gravity
ζ = drag coefficient, defined by Eq. (5)

For a steady motion the summ of this forces becomes zero:

$$\Sigma F = 0 = F_d + F_b - F_g \tag{6}$$

Equation (6) is the fundamental relationship for the calculation of settling rates.

The lines around a falling sphere in Fig. 6 are stream lines. Furthermore, Fig. 6 demonstrates the three mentioned forces — F_d as a surface force, F_g and F_b as volume forces.

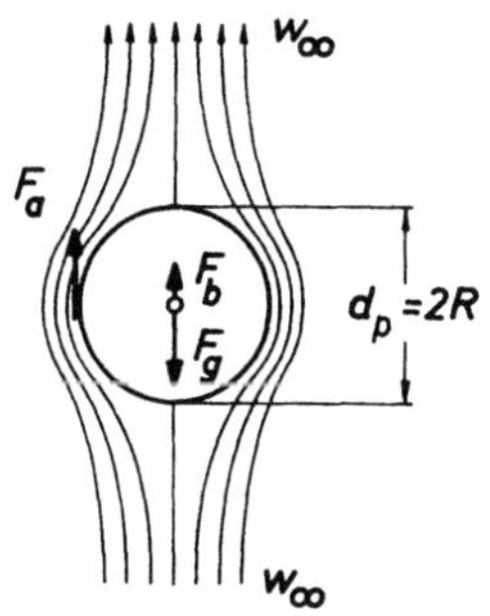

Fig. 6. Flow in contact with a spherical particle

For a reference system moving with the settling rate of the sphere it follows that

$$w_{s0} = w_\infty \tag{7}$$

where w_∞ = velocity of flow in the undisturbed liquid.
The drag coefficient is a function of w_{s0} or w_∞ respectively. If this function is known, one can calculate w_{s0} after substitution of Eqs. (3) ... (5) in Eq. (6).

Three ranges of flow characterized by the Reynolds number

$$Re = \frac{w_{s0} d_p}{\nu} \tag{8}$$

should be distinguished.

a) *Range with frictional forces dominating* Re < 1 *(creeping flow)*

By solving the momentum balance equation for the fluid flowing around a sphere, Stokes [1] calculated for the drag force

$$F_d = 6\pi\eta w_\infty R \tag{9}$$

where $R = d_p/2$ = radius of the sphere
$\eta = \nu\varrho$ = dynamic viscosity
ν = kinematic viscosity

From Eqs. (5), (8) and (9)

$$\zeta = \frac{24}{Re} \tag{10}$$

follows as a dimensionless form of Stoke's law. Introducing Eqs. (3), (4) and (9) into Eq. (6) and considering Eq. (7), gives for the settling rate

$$w_{s0} = \frac{4}{18} R^2 \frac{g}{\nu} \left(\frac{\varrho_p}{\varrho} - 1 \right) \tag{11}$$

b) *Range with frictional and inertial forces*, $1 < Re < 10^3$

Until now no theoretical solution has been obtained which is valid for the whole range mentioned.

By correlation of experimental results an approximate relationship [2] is given by:

$$\zeta = \frac{24}{Re} + \frac{4}{\sqrt{Re}} + 0{,}4 \tag{12}$$

Equation (12) is valid for $Re < 10^5$.

From Eqs. (12) and (5) the drag force can be obtained:

$$F_d = 6\pi\eta w_{s0} R + \frac{\varrho w_{s0}^2 \pi R^2}{2} \left(\frac{4}{\sqrt{Re}} + 0{,}4 \right) \tag{13}$$

Combining Eqs. (13), (8) and (6), the settling rate w_{s0} can be again calculated, but it cannot be given in an explicit form such as Eq. (11).

c) *Range with dominating inertial forces*, $10^3 < Re < 10^5$

From Eq. (12)

$$\zeta \approx 0.4 \tag{14}$$

follows (Newton's law).

Here the drag force is no longer a function of viscosity:

$$F_d = \frac{\varrho w_{s0}^2 \pi R^2}{5} \tag{15}$$

Introducing Eqs. (3), (4) and (15) into Eq. (6) the settling rate becomes

$$w_{s0} = \sqrt{\frac{20}{3} Rg \left(\frac{\varrho_p}{\varrho} - 1 \right)} \tag{16}$$

Figure 7 is a plot of the drag coefficient ζ versus Re-number according to Eq. (12). It contains all information with respect to the drag force for a newtonian fluid flowing around a sphere.

Figure 8 demonstrates the dependence of the settling rate w_{s0} of a spherical particle in water for a temperature of 10 °C from the particle diameter d_p and from the density relationship ϱ_p/ϱ. For other temperatures and other fluids, different plots are obtained.

Single microorganisms, flocs and particles formed by mycelia settle at a relatively low rate in the range of creeping flow ($Re < 1$) because of the small diameters and/or

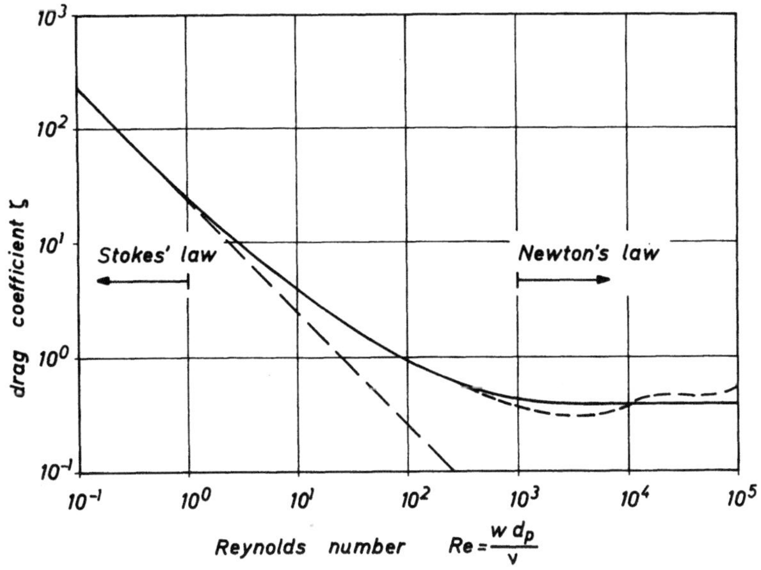

Fig. 7. Drag coefficient versus Reynolds number for a flow around a spherical particle

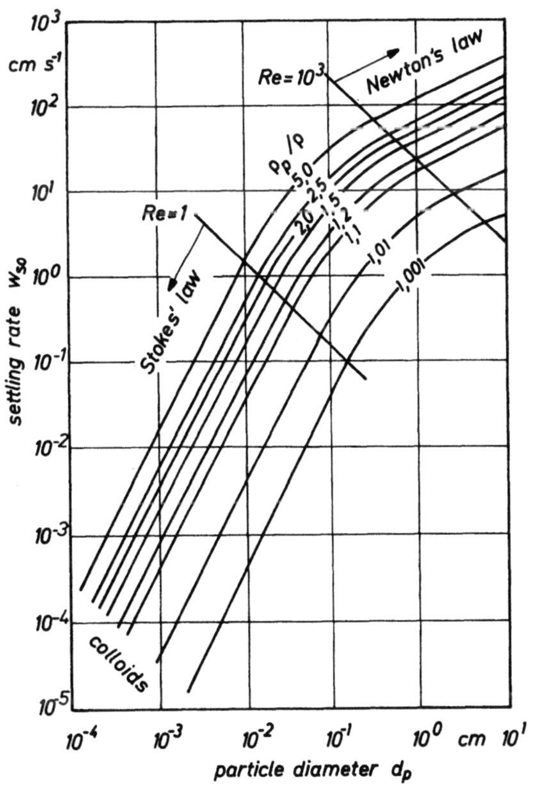

Fig. 8. Settling rates of single spherical particles in water for 10 °C (from Ref. [64])

the low differences in density. For known diameters and densities the settling rate can also be approximately calculated from Eq. (11), if the particle is not spherical.

2.2.2 Sedimentation in the Range of High Particle Concentration

2.2.2.1 Thickening of Ideal Suspensions

If a settling glass is filled for example with a relatively highly concentrated $CaCO_3$-suspension, certain remarkable properties of the system can be observed after mixing and during sedimentation [3] (Fig. 9):

— After 5 min a *clear liquor zone A* is formed with only some small suspended particles so that free settling can occur according to the principles outlined of chapter 2.2.1. Below it a zone with an approximately constant local concentration can be seen the *zone of collective subsidence* B [4]. "The larger particles serve to collect and strain out the finer particles. The concentration of solids is so large that each particle substantially retains its position with respect to all of its neighbouring particles. Therefore, this zone has the concentration of the original slurry" [4].

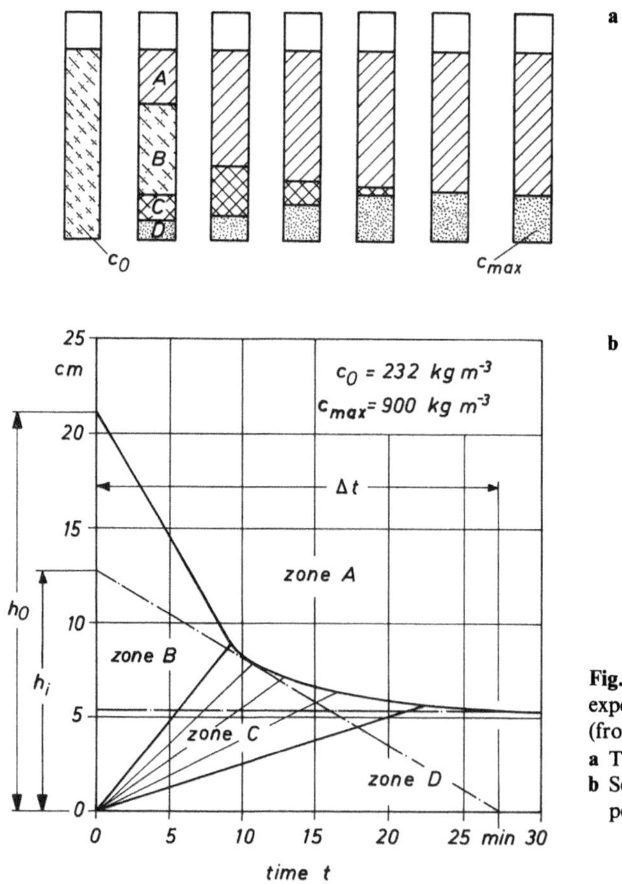

Fig. 9a and b. Batch sedimentation experiments with $CaCO_3$ slurries (from Ref. [3]).
a The four zones of sedimentation;
b Settling curve of an ideal suspension

At the bottom, the *compaction zone D* is formed by settled particles. Between zones B and D the *compression zone C* appears in which the settling rate decreases from the initial value to zero.

- After 10 min zone B with only a small depth can be observed. Compared to this zone D has been grown considerably.
- After 15 min zone B has disappeared.
- After 25 min all particles have reached zone D which retains a constant height and a maximum concentration during the following time.

The behaviour of the suspension in zone B is very interesting since all particles with different sizes and densities settle at approximately the same rate. The principles of "free settling" as described in 2.2.1 are no longer valid. Therefore, this type of settling is termed *zone sedimentation*. The settling rate w_s of this zone B of collective subsidence or zone sedimentation can be calculated from the diminution of the boundary h between zones A and B at time t and follows from the constant slope of the function h = h(t) (Fig. 9):

$$w_s = -\frac{dh}{dt} \tag{17}$$

If experiments are carried out with higher concentrations, settling curves with lower slopes can be obtained: the settling rate decreases with increasing solid concentration!

However this method for measuring settling rates can be only used,

- if the concentration is high enough, so that a zone of collective subsidence will arise
- and if the concentration is not too high, so that the continuous phase is still formed by the liquid and the disperse phase by the solid.

Some results for the sedimentation of $CaCO_3$ particles were plotted in Fig. 10. It shows that the dependence of the settling rate on the solid concentration can be described approximately by

$$\frac{w_s}{w_{s0}} = \left(1 - \frac{c}{c_{max}}\right)^m \tag{18}$$

with m = 4.36. Equation (18) was first mentioned by Richardson and Zaki [5]; they found an exponent of m = 4.65 for the sedimentation of equal glass spheres, w_{s0} is the settling rate of single particles and c_{max} the maximum concentration of zone D. In Eq. (18) c and c_{max} can be used as volume or mass concentration. Functions describing the dependence of settling rate from solid concentration, will be called subsequently "settling characteristics".

If suspensions show the discussed thickening behaviour and if the settling rate is only a function of solid concentration, they will be called "ideal suspensions". Although results of these batch experiments in settling glasses and the method of evaluation was allready published in 1916 by Coe and Clevenger [6], a mathematic analysis was first derived by Kynch [7] in 1952. According to his theory, zone C with concentrations increasing from the initial value c_0 to the maximum value c_{max} arises from the upwards moving boundary of the zone D against the settling particles.

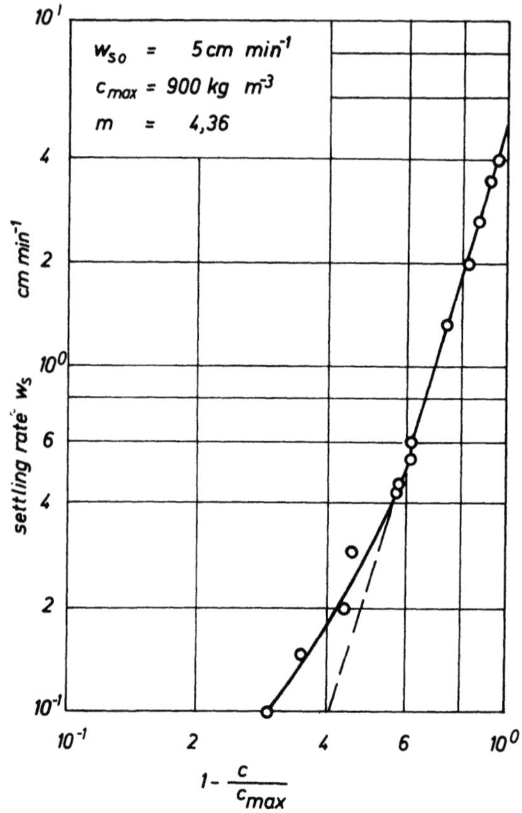

Fig. 10. Settling rate versus dimensionless solid concentration $\left(1 - \frac{c}{c_{max}}\right)$ for sedimentation of $CaCO_3$ particles in water (from Ref. [3])

Kynch pointed out that the settling rate w_{si} of the boundary between the zones A and C has to be

$$w_{si} = \frac{h_i}{\Delta t} \tag{19}$$

assigned to a concentration c_i, which can be calculated by

$$c_i = \frac{c_0 h_0}{h_i} \tag{20}$$

where h_0 = initial height of the suspension
h_i = intercept of the ordinate formed by the tangent to any point of the settling curve and the origin of coordinates (Fig. 9)
Δt = intercept of the abscissa formed by the tangent and the origin of coordinates (Fig. 9).

Therefore the whole settling characteristic can be obtained from only one settling curve by the construction of different tangents to different points of the curved part of the settling curve. However, for this method a very high accuracy of measurement is necessary. More exact results can be expected by the measurement of different settling curves for different initial solid concentrations.

The most important assumption for this evaluation of batch settling experiments is the unique dependence of settling rate on solid concentration. This assumption is approximately valid for rigid particles like $CaCO_3$ crystals. However, for flocs considerable deviation from the discussed properties of the two phase system can be observed.

2.2.2.2 *Thickening of "Non-ideal Suspensions"*

The additional problems in describing phenomena settling in non-ideal suspensions are discussed for activated sludge flocs. The first difficulties will appear, if we want to define the geometric shape of these flocculent particles. Because of their low resistance to deformation their shapes and sizes are particulary changed in regions with high velocity gradients. The considerable high water content consists of three portions:

— The *floc water* is hindered extensively from flowing through the particle by the fibrous structure of the solid substance. Flocs, which have reached the zone D in batch settling tests, contain all this floc water only in the first moment. Then it is pressed out increasingly, simultaneously the flocs are compacted. In centrifuges a centrifugal acceleration number of Z = 3000 (Sect. 3.2.1) is necessary to press floc water out completely (Fig. 11).

— To separate the *capillary water*, centrifugal acceleration numbers of Z = 15000 are needed which can be reached only in small-scale laboratory centrifuges.

— The *particle water* is adsorbed at the solid surfaces and enclosed in microorganisms. It can be removed only by drying after lysis.

As can be seen in Fig. 11, the floc water content of activated sludge flocs predominates the residue particle volume consisting of capillary water, particle water and solid

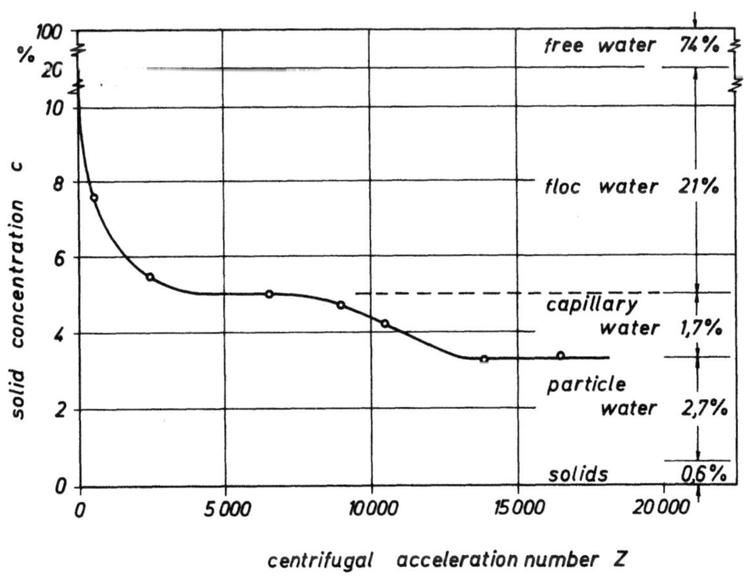

Fig. 11. Classification of water in a waste activated sludge by gravitational and centrifugal settling (from Ref. [9])

material by a factor four. For this example a floc density of $\varrho_p = 1.024 \text{ g cm}^{-3}$ can be calculated which is only insignificantly higher than the density of water.

Activated sludge reactors contain frequently solid mass concentrations of 4 g l^{-1} as dry matter. With a floc density of $\varrho_p = 1.024 \text{ g cm}^{-3}$ a mass concentration of $c = 122 \text{ g l}^{-1}$ follows, which is 12.2% by volume! This concentration is usually increased by thickening in the secondary settling tank of an activated sludge plant by a factor of 3 ... 5.

Because of their high water content and their dewatering properties flocs show some anomalies in batch settling experiments.

1. *Compaction of the settled flocs in zone D*

In contrary to settled particles from ideal suspension, flocs are compacted in zone D, during which *floc water* flows upwards. This compaction phase often ends only after several days. Figure 12 is a qualitative representation of a settling curve with a decreasing boundary between the zones D and A. Since the compaction of zone D influences settling rates in zone C the method of Kynch [7] for the determination of different settling rates from only one settling curve is no longer applicable.

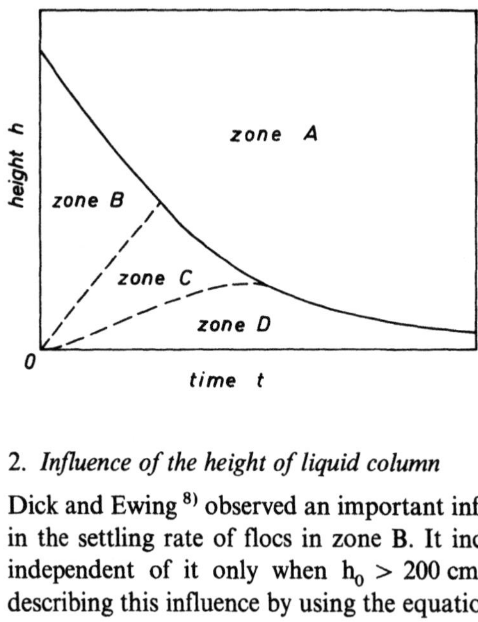

Fig. 12. Settling curve of non-ideal suspension

2. *Influence of the height of liquid column*

Dick and Ewing [8] observed an important influence of the height of liquid column h_0 in the settling rate of flocs in zone B. It increases with increasing h_0 and becomes independent of it only when $h_0 > 200$ cm (Fig. 13). The authors succeeded in describing this influence by using the equation

$$w_{sh} = \frac{h_0}{a + \frac{h_0}{w_s}} \tag{21}$$

By plotting h_0/w_{sh} versus h_0 straight lines were obtained (Fig. 14) and the coefficient a as well as w_s are obtained from the slope and the intercept on the ordinate. The coefficient a is called "*retardation factor*". It gives information about

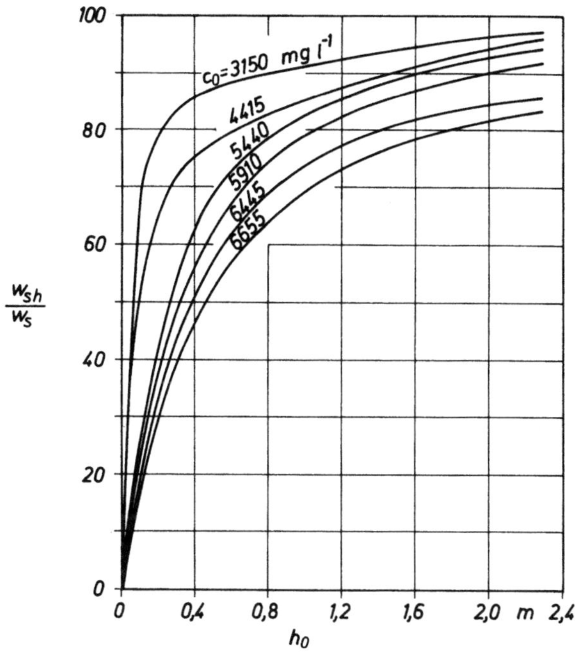

Fig. 13. Dimensionless settling rate w_{sh}/w_s versus height of the fluid column h_0 for an activated sludge (from Ref. [8]), parameter: initial solid concentration c_0

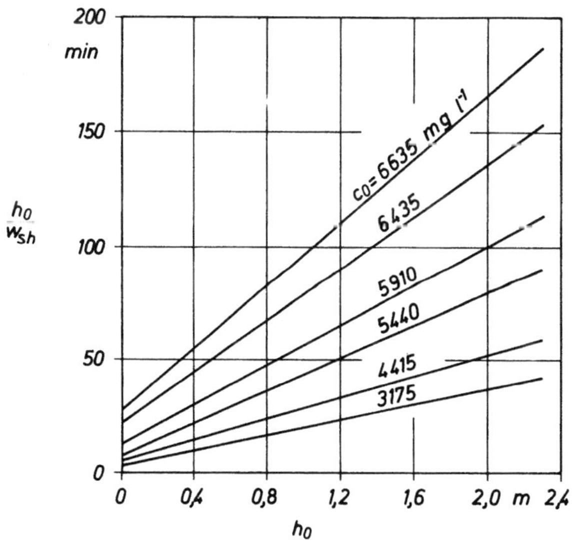

Fig. 14. h_0/w_{sh} versus h_0 (see Fig. 13)

the divergence from the behaviour of ideal suspensions. For the sedimentation of silica, a retardation factor $a = 0$ was found, whereas for bulking sludge with a high content of filamentous bacteria, a particularily high retardation factor was obtained.

3. *Influence of cylinder diameter*

This influence on the settling rate of activated sludge in zone B was investigated by Veselind [9]. The settling rate was completely independent of the diameter d only when

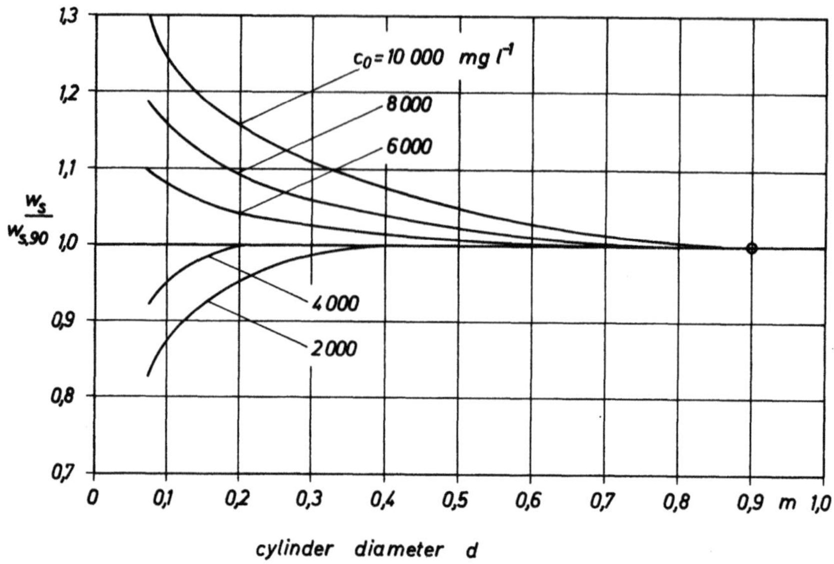

Fig. 15. Dimensionless settling rate $w_s/w_{s,90}$ versus cylinder diameter d for an activated sludge (from Ref. [9]), parameter: initial solid concentration c_0
$w_{s,90}$ = settling rate in a cylinder with a diameter d = 90 cm

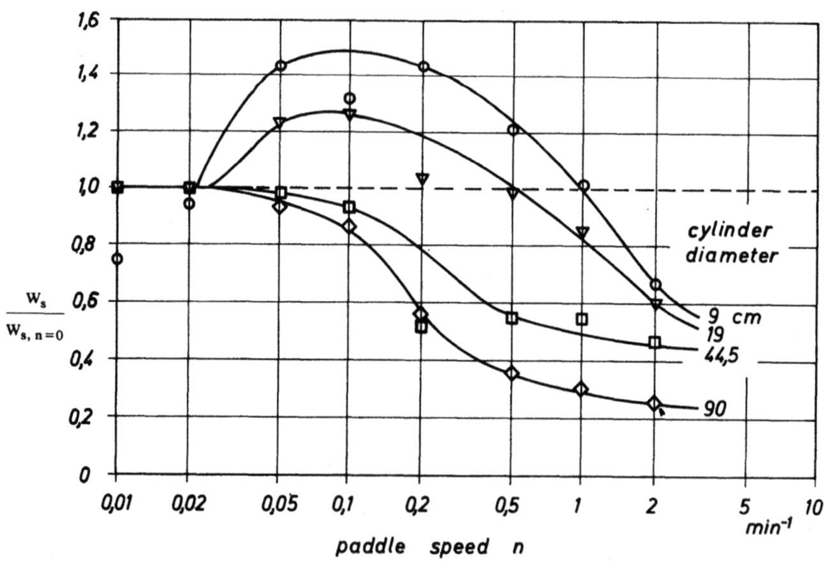

Fig. 16. Dimensionless settling rate $w_s/w_{s,n=0}$ versus speed of rotation n of paddle stirrer (from Ref. [9]).
parameter: cylinder diameter d;
$w_{s,n=0}$ = settling rate for n = 0;
— effect of the intensity of a mixing process before batch settling tests —

d ≧ 90 cm. For smaller diameters lower settling rates were obtained for mass concentrations of 2 and 4 g l^{-1}, whereas higher rates follow from experiments with concentrations of 6 to 10 g l^{-1} (Fig. 15). The author recommended to use cylinder diameters of at least 20 cm for activated sludge settling tests.

4. *Influence of mixing intensities before settling tests*

A homogeneous suspension is often produced by stirring. Figure 16 represents the influence of the speed of rotation on settling rates for four different cylinder diameters between 9 and 90 cm [9]. For lower rotational speeds and smaller diameters the settling rate increases probably as a consequence of flocculation effects. At higher rotational speeds and with larger diameters, lower settling rates were observed because of the destruction of larger flocs. For higher diameters, destruction was obviously more effective then flocculation.

Therefore, the following prerequisites should be fulfilled, if the investigator wishes to measure correct settling rates of flocs:

1) Because of floc compaction in zone D, for each solid concentration an additional settling test must be carried out.
2) The height of the liquid column has to agree with the depth of the technical settling tank or experiments with different heights are necessary to find operational conditions without influences of column height.
3) The cylinder diameter of the settling glass must be at least 20 cm.
4) If a stirring process is necessary for homogenization of the suspension, the type of the stirrer and the rotational speed will have to be carefully selected, in order to reduce flocculation and destruction effects.

Table 1 shows some settling characteristics for different biosuspensions. In the case of activated sludge an exponential equation is frequently used. However it is doubtful,

Table 1. Settling characteristics of biosuspensions

Biosuspension	Settling characteristic w_s in mh^{-1}, c in g l^{-1} (dry matter)	k	m	Ref.	Year
Conventional activated sludge	$w_s = kc^{-m}$	22,8	2,5	10)	1957
Conventional activated sludge	$w_s = kc^{-m}$	42,0	1,63	8)	1967
Conventional activated sludge	$w_s = kc^{-m}$	5,6	2,62	11)	1970
Pure oxygen activated sludge (11 plants)	$w_s = kc^{-m}$	3,0 ... 27,8	2,25	12)	1974
Conventional activated sludge	$w_s = kc^{-m}$	6,0	1,70	13)	1974
Conventional activated sludge	$w_s = 14 \exp(-0{,}92c)$			14)	1981
Conventional activated sludge	$w_s = 5{,}2 \exp(-0{,}38c)$			15)	1979
Saccharomyces cerevisiae	$w_s = \frac{1}{1 + 5c'^{1/3}}$; c' = volume concentration; $c' = 0{,}2 \times 10^{-1}\,c$			16)	1979

whether or not the previously discussed prerequisites were considered in all investigations, so that additional problems can arise in scale-up procedures for continuous thickening processes based on laboratory batch settling test data (Sect. 2.3.2).

2.3 Sedimentation in Open Systems

2.3.1 Clarification

2.3.1.1 Model for Calculating Clarification Ratios from Suspensions of Equallysized Particles

The fundamental principle of modelling clarification processes is the calculation of particle trajectories on the basis of hydrodynamic considerations.

The model used here is based on the following assumptions:

1) The settling rates of all particles correspond to each other and are constant for all locations and times.
2) Fluid flow is laminar.
3) Particle concentration is constant in the inlet cross sectional area.

The consequences are to be discussed for the example of a rectangular horizontal flow tank (Fig. 17):

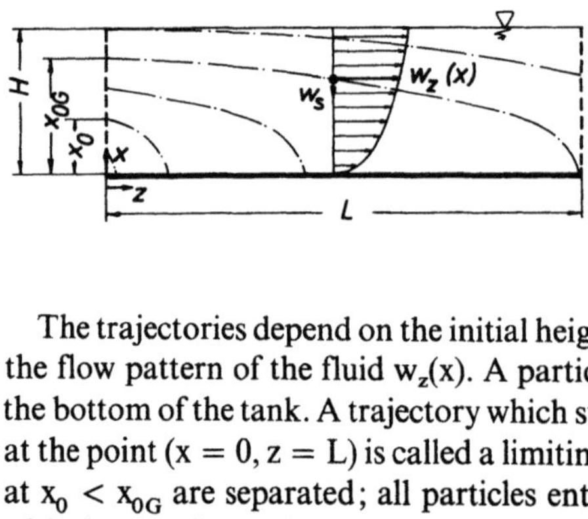

Fig. 17. Particle trajectories in a rectangular horizontal flow tank

The trajectories depend on the initial height of the particle x_0, the settling rate w_s and the flow pattern of the fluid $w_z(x)$. A particle will be separated if a trajectory ends at the bottom of the tank. A trajectory which starts at the point ($x = x_{0G}, z = 0$) and ends at the point ($x = 0, z = L$) is called a limiting trajectory. All particles entering the tank at $x_0 < x_{0G}$ are separated; all particles entering the tank at $x_0 > x_{0G}$ are discharged with the overflow. Therefore, the clarification ratio can be calculated from

$$\beta = \frac{\int_0^{x_{0G}} w_z(x)\,dx}{\int_0^{H} w_z(x)\,dx} \tag{22}$$

where H = height of liquid in the tank.

The open systems now discussed differ from other forms of the limiting trajectory and other flow patterns.

2.3.1.2 Rectangular Horizontal Flow Tank

The flow pattern of an open channel with a rectangular cross section area is given by

$$w_z(x) = \frac{3}{2}\left[2\frac{x}{H} - \left(\frac{x}{H}\right)^2\right] w_0 . \tag{23}$$

After introducing Eq. (23) in Eq. (22) and integrating it follows that

$$\beta = \frac{3}{2}\left(\frac{x_{0G}}{H}\right)^2 - \frac{1}{3}\left(\frac{x_{0G}}{H}\right)^3 . \tag{24}$$

In Eq. (24) the initial height of the limiting trajectory x_{0G} is still unknown. By solving the differential equation of the trajectories

$$w_z(x)\, dx = - w_s dz \tag{25}$$

and by considering the boundary conditions

$$\begin{aligned} x &= x_{0G} \quad \text{for} \quad z = 0 \\ x &= 0 \quad \text{for} \quad z = L \end{aligned} \tag{26}$$

x_{0G} can be calculated, so that Eq. (24) now becomes

$$\beta = \frac{L\, w_s}{H\, w_0} . \tag{27}$$

This result was already given by Hazen [17] in 1904. It is valid for all kinds of laminar flow patterns, even for additional functions from the coordinate in flow direction x

$$w_z = w_z(z, x) \tag{28}$$

which follows from flow conditions near the inlet and exit part of the tank (Wouda, Rietema and Ottengraf [18], Fig. 2a). However Eq. (27) is not valid for flow patterns with a significant influence of the horizontal coordinate y perpendicular to flow direction [19].

For the limit of a complete particle separation ($\beta = 1$) it follows from Eq. (27)

$$A = \frac{\dot{V}_0}{w_0} , \tag{29}$$

where $A = BL$ = tank surface
B = tank width
and $\dot{V}_0 = w_0 HB$ = flow rate.

In the form of Eq. (29) the result is often used for the dimensioning of rectangular horizontal flow tanks.

Under operation conditions several disturbing influences caused by wind, density differences, turbulent flow and sludge scrapers occur which often results in lower clarification ratios. But even in laboratory measurements remarkable deviations from the assumed flow pattern can be observed (Camp [20]) (Fig. 18).

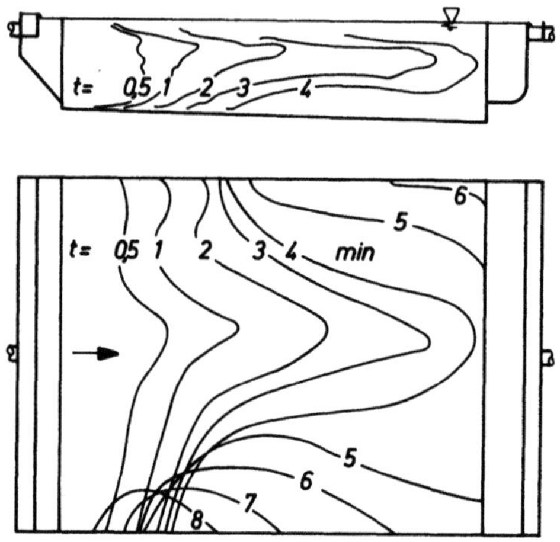

Fig. 18. Concentration fronts in a bench scale rectangular horizontal flow tank, which were visualized by using coloring agents (from Ref. [20])

2.3.1.3 Circular Horizontal Flow Tanks

Figure 19 shows a vertical cross section and a limiting trajectory. Instead of the coordinate z the cylindrical coordinate r must be used. The fluid velocity w_r decreases because of the increasing cross section areas. Hence for laminar flow, the flow pattern is given by

$$w_r(x, r) = \frac{3}{2}\left[2\frac{x}{H} - \left(\frac{x}{H}\right)^2\right] w_{02}\frac{r_2}{r} \tag{30}$$

where w_{02} = mean velocity at $r = r_2$.

By using the same method as in Section 2.3.1.2 the clarification ratio can be calculated to be

$$\beta = \frac{r_2^2 - r_1^2}{2r_2H}\frac{w_s}{w_{02}}. \tag{31}$$

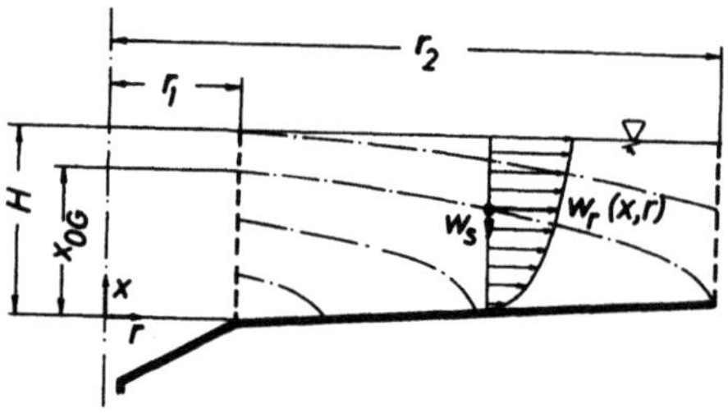

Fig. 19. Particle trajectories in a circular horizontal flow tank

2.3.1.4 Circular or Square Vertical Flow Tanks

In this case clarification occurs during upward flow and consequently in the upper part of the tank. All particles with settling rates $w_s > w_x$ will settle with the relative velocity $w_s - w_x$, particles with $w_s < w_x$ will be removed with the overflow. For floc suspensions however a sludge blanket can arise as a fluidized bed with relative high particle concentration since settling rate and fluid velocity coincide ($w_s = w_x$). Under these conditions high clarification ratios can be attained by additional filtration and flocculation effects which cannot be determined by sedimentation models.

2.3.1.5 Lamella Type of Separators

The method of calculation is demonstrated for an element of two inclined parallel plates and for countercurrent operation (Fig. 5b):
With the equation of the flow pattern

$$w_z(x) = 6\left[\frac{x}{H} - \left(\frac{x}{H}\right)^2\right] w_0 \tag{32}$$

it follows for the clarification ratio from Eq. (22) that

$$\beta = 3\left(\frac{x_{0G}}{H}\right)^2 - 2\left(\frac{x_{0G}}{H}\right)^3 . \tag{33}$$

From the differential equation of the trajectory (Fig. 20)

$$[w_z(x) - w_s \sin\alpha]\, dx = - w_s \cos\alpha\, dz \tag{34}$$

and the boundary conditions for the limiting trajectory (Eq. (26)) an implicit equation for $x_{0,G}$ can be given by

$$3\left(\frac{x_{0G}}{H}\right)^2 - 2\left(\frac{x_{0G}}{H}\right)^3 = \frac{x_{0G}}{H}\frac{w_s}{w_0}\sin\alpha + \frac{L}{H}\frac{w_s}{w_0}\cos\alpha \tag{35}$$

which can be solved by simple iteration methods [21, 23]. In contrary to the open channel problem (Sect. 2.3.1.2) different solutions are obtained for different flow patterns. However, Binder [19] indicated that these differences can be neglected for laminar and plug flow patterns. Therefore, the simple solution for plug flow can be used

$$\beta = \frac{\frac{L}{H}\cos\alpha}{\frac{w_0}{w_s} + \sin\alpha} . \tag{36}$$

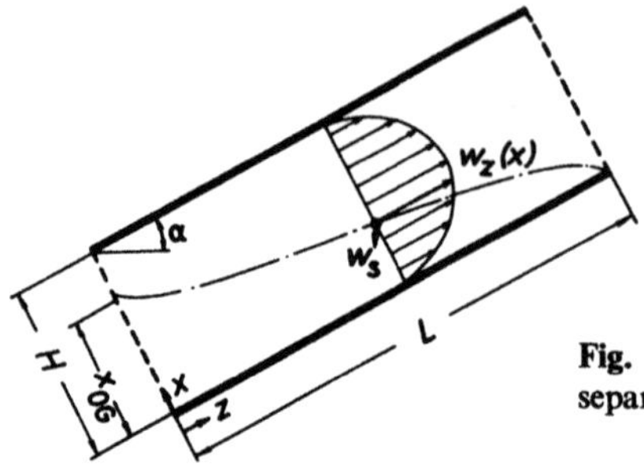

Fig. 20. Particle trajectories in a counter current lamella separator

For $\frac{L}{H \tan \alpha} \rightarrow \infty$ both solutions coincide. The condition for complete separation follows from Eq. (26) for $\beta = 1$

$$\frac{w_s}{w_0} = \frac{1}{\frac{L}{H} \cos \alpha + \sin \alpha} \tag{37}$$

Table 2 shows the adequate solutions for direct current and cross current.

With the same hydrodynamic model Binder obtained solutions for other types of lamella cross section areas as circles (tube settler) and rectangles with different ratios of sides [19, 23].

When compared with horizontal and vertical flow tanks, the flow in lamella separators is a small space flow in a laminar flow region. Therefore a good correlation with experimental results can be expected for suspensions with equally sized particles. This could be confirmed with anorganic materials in parallel plate and tube separators [19, 23] and with the yeast *Saccharomyces cerevisiae* in tube separators [16, 24,

Table 2. Clarification ratios and conditions for complete separation for lamella separators consisting of inclined parallel plates and different methods of operation [19]

Methods of operation	Clarification ratio	Condition for complete separation
Counter current	$\beta = \frac{\frac{L}{H} \cos \alpha}{\frac{w_0}{w_s} + \sin \alpha}$	$\frac{w_s}{w_0} = \frac{1}{\frac{L}{H} \cos \alpha + \sin \alpha}$
Direct current	$\beta = \frac{\frac{L}{H} \cos \alpha}{\frac{w_0}{w_s} - \sin \alpha}$	$\frac{w_s}{w_0} = \frac{1}{\frac{L}{H} \cos \alpha - \sin \alpha}$
Cross current	$\beta = \frac{B}{H} \frac{w_s}{w_0} \cos \alpha$	$\frac{w_s}{w_0} = \frac{1}{\frac{B}{H} \cos \alpha}$

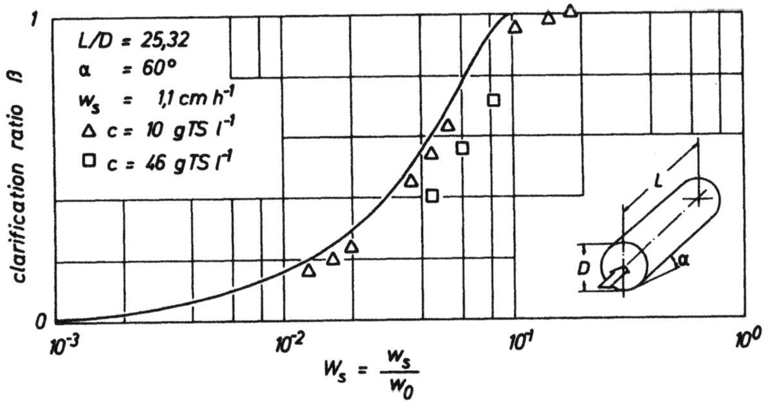

Fig. 21. Clarification ratio versus the coefficient w_s/w_0 for sedimentation of yeasts cells in an inclined tube clarifier (from Ref. [24])

[25]. Figure 21 shows some experimental and theoretical results for two different solid concentrations and at a settling rate $w_s = 1.1$ cm h^{-1} for yeasts with a diameter of 7 µm.

2.3.1.6 Model for Calculating Clarification Ratios of Suspensions of Polysized Particles

For calculating clarification ratios in systems with polysized particles the function of size distribution is needed. Because of the tendency of biosuspensions to flocculate and the low resistance of flocs to shear stress, it is difficult to obtain correct results by usual methods of size analysis.

Figure 22 is a qualitative plot of the frequency distributions versus particle diameter for the feed suspension and the underflow. The area formed by the frequency distribution curve of the feed and the abscissa is equal to the particle mass flow rate of the feed q_{30}, the shaded area below the frequency distribution curve of the settled particles is equal to the particle mass flow rate of the underflow $g'q_{3D}$. Hence, the clarification ratio of each particle size follows from the ratio of the two frequency distribution functions

$$\beta(d_p) = \frac{g'q_{3D}(d_p)}{q_{30}(d_p)} \tag{38}$$

For known β-values the clarification ratio of a polysized suspension β_Σ can be calculated from the expression

$$\beta_\Sigma = \int_0^{d_{p\,max}} \beta(d_p)\, q_{30}(d_p)\, dd_p\,. \tag{39}$$

Particles with $d_p > d_{pG}$ are separated completely, whereas particles with $d_p < d_{pG}$ are separated only in a fraction, which is given by $\beta(d_p < d_{pG})$. Therefore the integral can be divided into two parts

$$\beta_\Sigma = \int_0^{d_{pG}} \beta(d_p)\, q_{30}(d_p)\, dd_p + \int_{d_{pG}}^{d_{p\,max}} q_{30}(d_p)\, dd_p \tag{40}$$

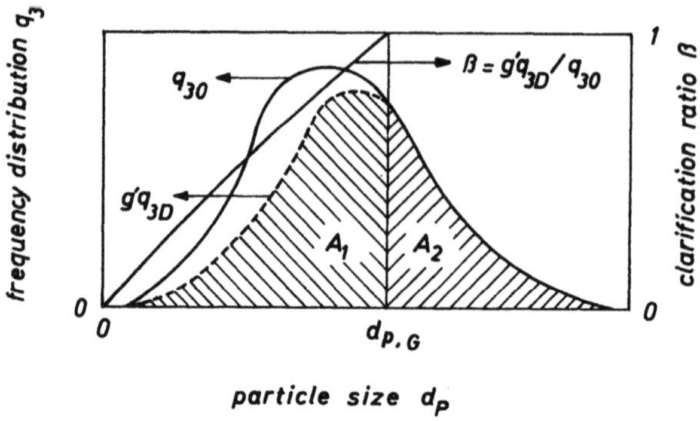

Fig. 22. Frequency distribution for the feed suspension g' q_{30}, the underflow q_{3D} and clarification ratio for monosized particles β versus particle diameter

with

$$Q_3(d_p) = \int_0^{d_p} q_3(d_p)\, dd_p \tag{41}$$

as the cumulative frequency distribution. Thus the universal relationship for the clarification ratio of polysized suspension can be expressed by

$$\beta_\Sigma = \int_0^{d_{pG}} \beta(d_p)\, q_{30}(d_p)\, dd_p + 1 - Q_3(d_{pG}) \tag{42}$$

For different types of separation apparatus, adequate functions for $\beta(d_p)$ must be used.

With the distribution function of Rosin-Rammler-Sperling-Bennet [26]

$$Q_3 = \exp\left(-\frac{d_p}{d'_p}\right)^n \tag{43}$$

some calculations were carried out [19,27]. In Fig. 23 the clarification ratio of a rectangular horizontal flow tank is plotted against the dimensionless coefficient Hz (Hazen number)

$$Hz = \frac{w_{s0.5} L}{w_0 H} \tag{44}$$

where $w_{s0.5}$ = settling rate of a particle diameter, belonging to $Q_3 = 0.5$.

The parameter is the exponent n of Eq. (43).

Figure 24 represents adequate results for an element of a countercurrent lamella separator consisting of two inclined parallel plates with the angle of inclination as a parameter. Results of measurements with silica powder ($d_p = 40\ \mu m$, $n = 1.25$) suspended in water are marked by points [19].

The results given in this chapter can be used for dimensioning sedimentation tanks if clarification is the most interesting process. If informations about the thickening

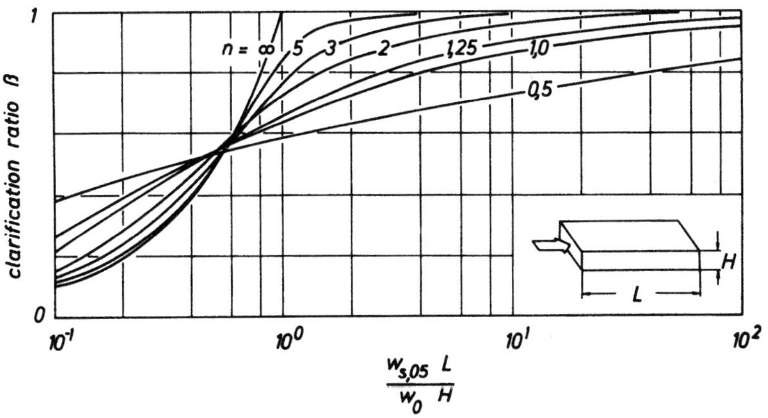

Fig. 23. Clarification ratio versus the coefficient $\frac{w_{s,05} \times L}{w_0 H}$ for sedimentation of polysized particles, in a parallel plate separator, following a distribution function of Rosin-Rammler-Sperling-Bennet (see Eq. (43))
parameter: distribution coefficient n

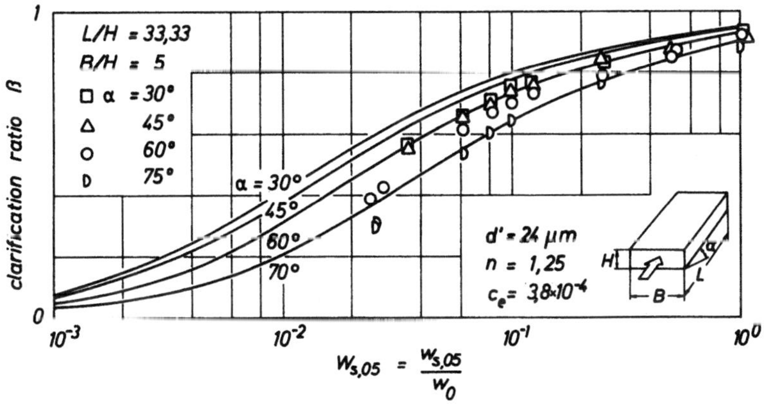

Fig. 24. Clarification ratio versus the coefficient $\frac{w_{s,05} L}{w_0 H}$ for sedimentation of polysized particles, in a parallel plate separator, following a distribution function of Rosin-Rammler-Sperling-Bennet (see Eq. (43))
parameter: angle of inclination
comparison of theoretical and experimental results

ratio of the sludge are of comparable significance, the thickening process has to be considered as well. The known methods are shown for a vertical flow tank as an example.

2.3.2 Clarification and Thickening in Vertical Flow Tanks

2.3.2.1 Conventional Scale-up Method Based on Sludge Volume Index and Operating Results

The Sludge Volume Index (SVI) in ml g^{-1} is the reciprocal solid concentration in the zone with settled particles obtained by a batch settling test in a sedimentation glass with a volume of 1 l and a diameter of 6.6 cm after a settling time of 30 min (Fig. 25). The SVI is used as a quantity to characterize the thickening qualities of an activated sludge.

The Comparison Sludge Volume (CSV) is defined by

$$\mathrm{CSV} = \mathrm{TS_R}\ \mathrm{SVI} \quad \text{in} \quad \frac{\mathrm{ml}}{\mathrm{l}} \tag{45}$$

where TS_R = dry solid matter of the undiluted sample of activated sludge in g l^{-1}.

In Fig. 26 the surface loading defined as

$$w_0 = \frac{\dot{V}_0}{A} \quad \text{in} \quad \frac{\mathrm{m}}{\mathrm{h}} \tag{46}$$

is plotted versus the Comparison Sludge Volume for horizontal (curve 1) and vertical flow tanks (curve 2) [28]. These curves are drawn by using the main values for different operating results. Table 3 shows some standard values of sludge volume indices for two different ranges of sludge loadings.

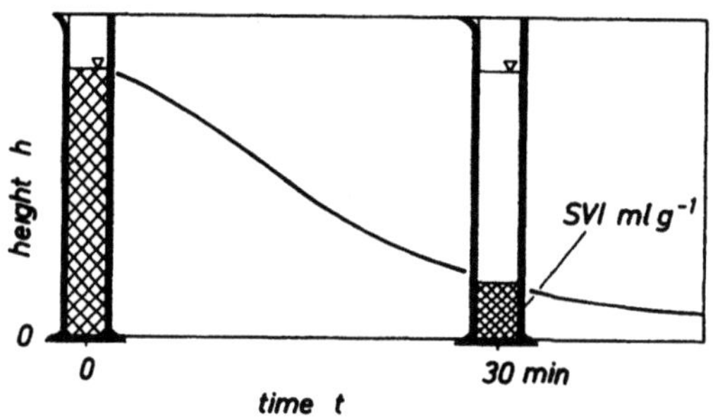

Fig. 25. Settling curve and Sludge Volume Index SVI

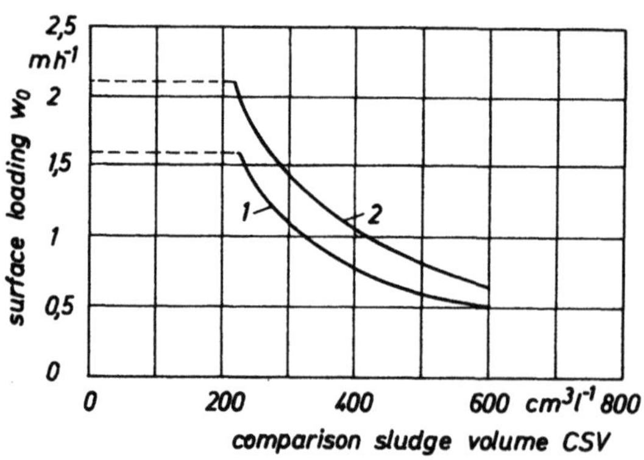

Fig. 26. Surface loading w_0 versus Comparison Sludge Volume CSV

After measuring or estimation of TS_R and SVI values, the comparing sludge volume is calculated. The surface loading of the secondary sedimentation tank can then be obtained from Fig. 26 so that the surface area of the tank A follows from Eq. (46) for a given flow rate $\dot{V}$ [29].

Table 3. Standard values of Sludge Volume Indices SVI in ml g^{-1} (Ref. [28])

Waste water	Sludge loading	
	$B_{TS} \geqq 0.05$	$B_{TS} \leqq 0.05$
	g BOD per g biomass day	
With a low part of organic industrial waste water	100–150	75–100
With a high part of organic industrial waste water	150–200	100–150

It is obvious that with SVI only one point of one settling curve is considered to characterize settling qualities [30,31]. In the following sections another method based on Kynch's theory is demonstrated. Since more laboratory settling data are considered, a more realistic scale-up can be expected. However, this method is truely valid only for cylindrical vertical flow tanks and ideal suspensions. Until now it has seldom been used in biological waste water treatment.

2.3.2.2 The Theory of Limiting Flux

Figure 27 is a schematic diagram of a cylindrical vertical flow tank and a typical concentration profile [3] with flow zones of different solid concentrations:

— the clear liquor zone A
— the zone of collective subsidence or sludge blanket zone B with a relative low constant concentration c_B,
— the compression zone C,
— and the compaction zone D.

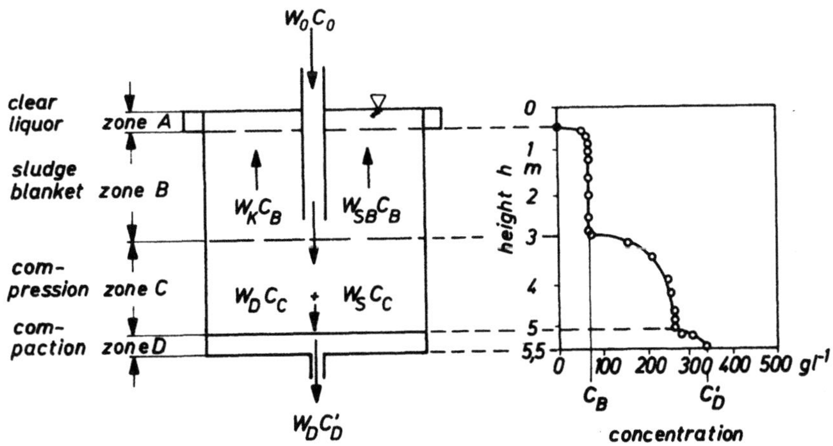

Fig. 27. Cylindrical vertical flow tank and a concentration profile for the thickening of a $CaCO_3$ slurry (from Ref. [3])

Solids and liquids are transferred with the following velocities:

- the settling rate $w_s = f(c)$
- the fluid velocity of the feed w_0
- the fluid velocity of the underflow w_D
- the fluid velocity of the overflow w_K.

In the compression zone C the total solids flux φ_Σ is the sum of the solids flux due to underflow $w_D c_C$ and the solids flux due to settling $w_s c_C$

$$\varphi_\Sigma = w_D c_C + w_s c_C \tag{47}$$

It is usual to use mass concentrations. Otherwise particle density measurements are needed.

Figure 28 is a plot of φ_Σ versus concentration for a settling characteristic given by Eq. (18). For a concentration c_L the total solid flux is equal to the minimum value.

The theory of the limiting flux can be formulated as follows: In the compression zone C a layer of concentration c_L can form and determines the maximum solids flux or limiting flux. c_L is a function of the fluid velocity of the underflow and the settling characteristic $w_s = w_s(c)$. Kynch [7] was the first person who described batch sedimentation of an ideal suspension. He showed that the straight lines in Fig. 9 going through the origin of coordinates are lines of constant solid concentrations. The upwardly directed travelling rate of these concentration layers w is given by the slope of the straight lines and it depends on the settling characteristic. Kynch demonstrated, that

$$w = -\frac{d(w_s(c)\, c)}{dc} \tag{48}$$

must be valid.

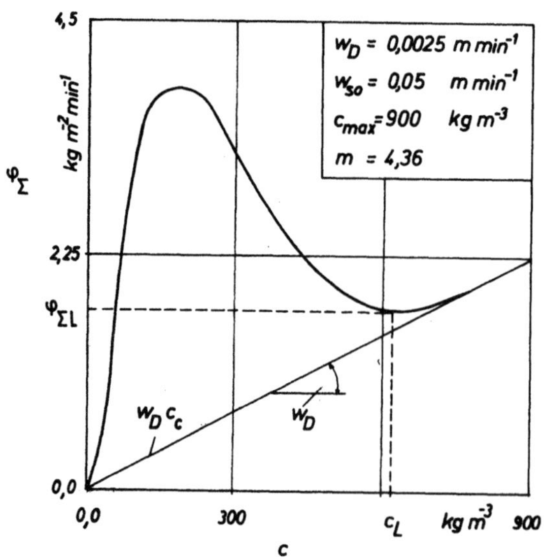

Fig. 28. Total solids flux φ_Σ and solids flux due to the fluid velocity of the underflow versus solid concentration, settling characteristic of Richardson and Zaki [5]:

$$\frac{w_s}{w_{s0}} = \left(1 - \frac{c}{c_{max}}\right)^m$$

Hence, for a continuous thickening process it can be concluded that if the underflow is drained off with a velocity

$$w_D < w$$

a portion of total solids flux will be transferred upwards with the velocity $w - w_D$ and finally lost with the overflow. The tank is overloaded. For

$$w_D = w$$

the maximum load is obviously reached for which just no solid is lost. Therefore the operating condition for "maximum load" is given by

$$w_D = -\frac{d(w_s(c)\,c)}{dc} \tag{49}$$

Equation (49) coincides with the criterion for an extreme value in Fig. 28.
For

$$w_D > w$$

no transport limiting solid layer can arise. The loading can be increased without losing solids. The tank is underloaded.

Figure 29a represents a tank within an operation range of low loading. It works with reserve capacity, since the total solids flux is smaller than the limiting flux $\varphi_{\Sigma L}$. With increasing load the operating condition "maximum load" is finally reached. The tank still works without losing solids, however, a reserve of loading capacity does not exist any longer (Fig. 29b). Figure 29c shows flux curves for an overloaded tank: the total solid flux is higher than the limiting flux, consequently solids are lost with the overflow.

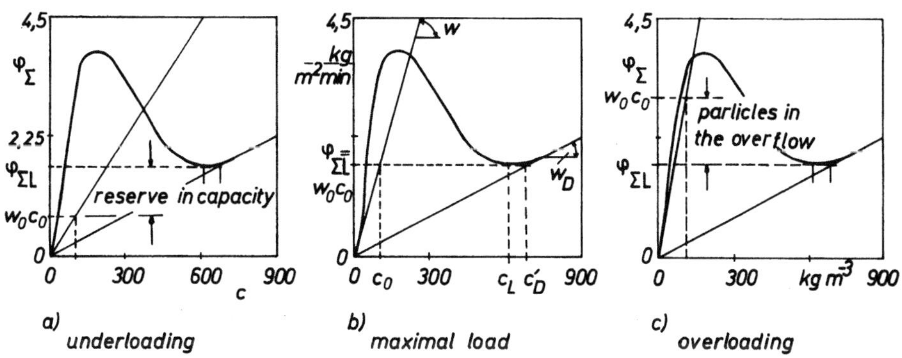

Fig. 29a—c. Total solid flux φ_Σ and solid flux due to the fluid velocity of the underflow — the three operating conditions —

2.3.2.3 *Material Balances*

In order to calculate the clarification ratio β and the thickening ratio n_E, the solid concentrations in the overflow c_k and in the underflow c'_D have to be known (see Eqs. (1) and (2)). They are functions of the fluid velocity of the feed w_0, of the fluid velocity of the underflow w_D, and of the solid concentration of the feed c_0 and derive from mass balance equations (Fig. 30):

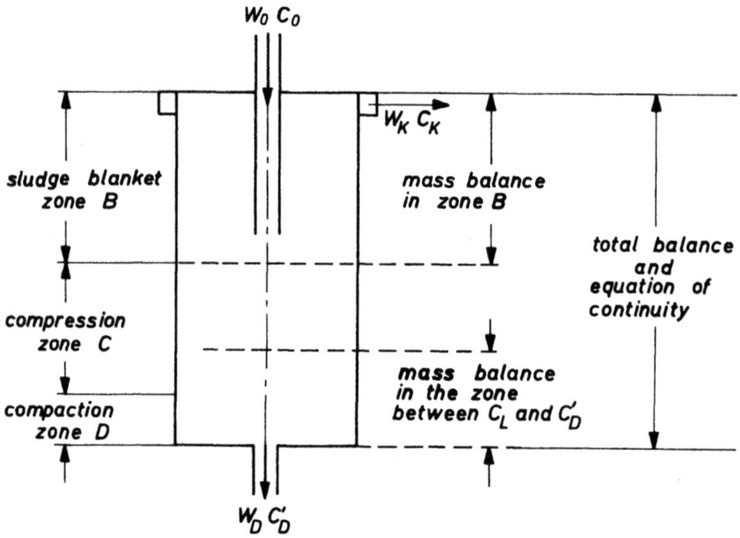

Fig. 30. Illustration of the theoretical model for continuous clarification and thickening in vertical flow tanks

Total mass balance is:

$$w_0c_0 = w_Dc'_D + w_Kc_K \tag{50}$$

and the continuity equation:

$$w_0 = w_D + w_K\,. \tag{51}$$

The mass balance in the zone between the layer with the concentration c_L and the lower boundary of zone D with concentration c'_D is:

$$w_Dc_L + w_s(c_L)\,c_L = w_Dc'_D\,. \tag{52}$$

For complete particle separation by clarification ($c_K = 0$) it follows from Eq. (50) that

$$w_D = \frac{w_0c_0}{c'_D} \tag{53}$$

and from Eqs. (50) and (52)

$$w_0c_0 = w_s(c_L)\,c_L + w_Dc_L\,. \tag{54}$$

By combining Eqs. (53) and (54), we obtain

$$w_0c_0 = \frac{w_s(c_L)}{\dfrac{1}{c_L} - \dfrac{1}{c'_D}}\,. \tag{55}$$

Equation (55) was published by Coe and Clevenger [6] in 1916. Indeed, they assumed a solid layer with the concentration c_L and a limiting flux $(w_s(c_L) + w_D)\,c_L$ however, they were only able to solve Eq. (55) approximately by a rough graphical method.

2.3.2.4 Graphical Solution by Yoshioka

Yoshioka [32] proposed the following graphical method:

- By using the settling characteristic, measured in batch settling tests, the solid flux due to settling $w_s(c)\,c$ is plotted versus solid concentration (Fig. 31).
- A second system of coordinates with the same abscissa but with the ordinate w_Dc is placed upon the first system in such a way that positive w_Dc-values show downwards and that both origins of coordinates exhibit a distance of w_0c_0.
- Starting from the origin of coordinates of the second system ($w_Dc = 0$, $c = 0$), a tangent to the curve $w_sc = f(c)$ is drawn, so that the abscissa of the point of contact is greater than that of the turning point W.

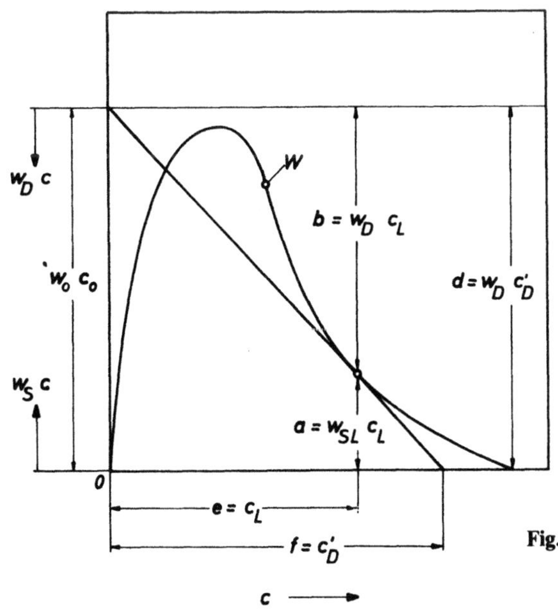

Fig. 31. Graphical solution by Yoshioka [32]

From Fig. 31 it can be concluded firstly that since the balance Eqs. (56) and (57) have to be satisfied, the lines a, b, d, e, f can be identified as

$$a = w_s(c_L)\,c_L$$

$$b = w_D c_L$$

$$d = w_D c'_D$$

$$e = c_L$$

and

$$f = c'_D\,.$$

Secondly, the slope of the tangent is $-\,w_D$. Therefore, at the point of contact Eq. (49) is satisfied and c_L can be interpreted as the concentration of the limiting layer.

Consequently c'_D can be determined by simple construction of tangents for different values of $c_0 w_0$ and w_D and for any given settling characteristic.

2.3.2.5 Analytical Solution

In order to optain an analytical solution, the settling characteristic must be given as a correlation function. In this case an exponential equation

$$w_s = kc^{-m} \tag{56}$$

is assumed which is frequently used to describe activated sludge settling datas (Table 1). Equation (56) is an approximation for a definite concentration range; it is not valid for very low and very high solid concentrations.

Figure 32 shows the results of settling measurements of activated sludge aerated in a pilot plant with pure oxygen (Dick and Young [33]), which follows Eq. (56). It is

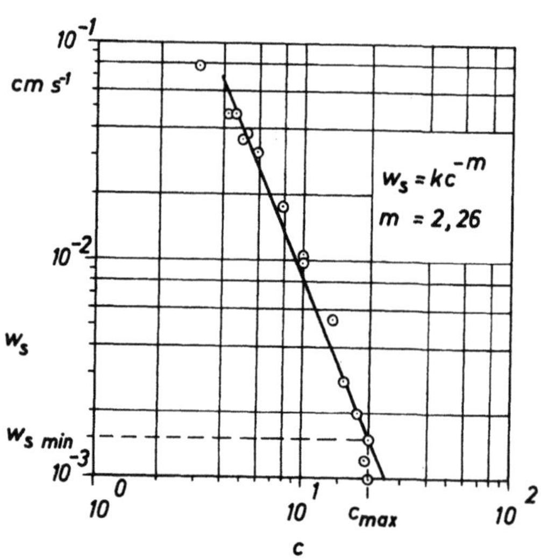

Fig. 32. Settling characteristic of activated sludge, aerated with pure oxygen in a pilot plant (from Ref. [33])

useful to make the concentration with c_{max} and all velocities with $w_{s\,min}$ dimensionless:

$$C = \frac{c}{c_{max}}$$

$$W_s = \frac{w_s}{w_{s\,min}} \tag{57}$$

$$W = \frac{w}{w_{s\,min}}$$

c_{max} is the maximum concentration of a suspension whose settling rate follows Eq. (56). $w_{s\,min}$ is the inherent minimum settling rate (Fig. 32). Then the problem can be expressed by the following Eqs.:
continuity equation:

$$W_0 = W_D + W_K \tag{58}$$

mass balances:

$$W_0C_0 = W_DC'_D + W_KC_K \tag{59}$$

$$W_DC'_D = W_DC_L + W_S(C_L)\,C_L \tag{60}$$

settling characteristic:

$$W_S = C^{-m} \tag{61}$$

operating condition for maximum load:

$$W_D = -\frac{d(W_S(c)\,C)}{dC} \tag{62}$$

For the proposed settling characteristic the equations can be analytically solved.

a) *The operating condition "underloading"*

In the case of a high underflow rate W_D, the transport limiting solid layer cannot form. Therefore, the load can be increased without loss of solids. To calculate C'_D only Eq. (59) is needed. For $C_K = 0$, C'_D is given by

$$C'_D = \frac{W_0C_0}{W_D} \tag{63}$$

In Fig. 33 C'_D is plotted against W_D for $C_0 = 0.2$ and $m = 2.26$ with W_0 as a parameter. For $W_0 = 20$ and $W_D = 20$ (point P1) all solids are drained off with the underflow (limiting case: hydraulic transport downward).

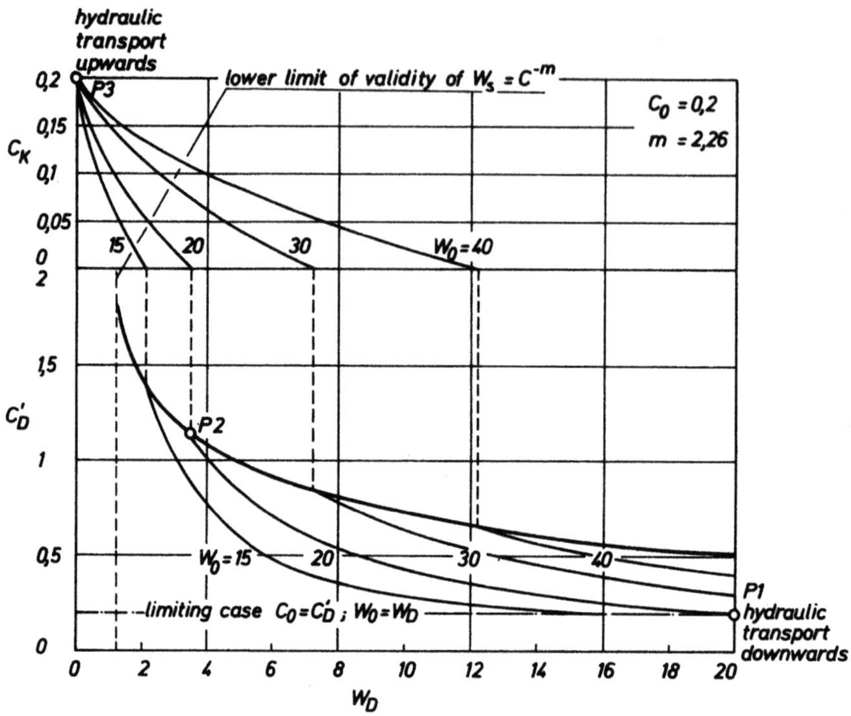

Fig. 33. Dimensionsless solid concentration in the underflow C_D and the overflow C_K versus dimensionsless fluid velocity due to underflow W_D
parameter: dimensionsless fluid velocity due to the feed W_0

b) *The operating condition "maximum load"*

If for $W_0 = 20$ the underflow rate W_D is reduced operating point P2 will be reached finally, which marks the operating conditions "maximum load". In this case with $C_D' = C_{DG}$ the highest thickening ratio n_E follows without loss of solids ($C_K = 0$, respectively $\beta = 1$).

To calculate C_{DG}, Eqs. (60), (61) and (62) are needed. The concentration of the transport limiting layer following from Eqs. (61) and (62) is

$$C_L = \left[\frac{m-1}{W_D}\right]^{1/m} \tag{64}$$

Combining Eqs. (64) and (60) we obtain

$$C_{DG} = W_D^{-1/m}[(m-1)^{1-m/m} + (m-1)^{1/m}] \tag{65}$$

All possible operating conditions "maximum load" are represented by Eq. (65), the accentuated curve in Fig. 33. However, it has to be assumed that C_L lies within the range of validity of Eq. (61), that is

$$0.2 \leqq C_L \leqq 1$$

(see Fig. 31). Then the corresponding limiting underflow rates follow from Eq. (64) to $W_{D,\,min} = 1.26$ and $W_{D,\,max} = 47.9$.

c) *The operating condition "overloading"*

If for $W_0 = 20$ the underflow rate W_D is reduced to a higher degree, the operating point will move on the curve "maximum load" (Eq. (65)) to the left. This results in higher solid underflow concentrations C_D', but simultaneously solids are lost with the overflow. Therefore, the operating point also moves on a second curve of an upper plot of Fig. 33, giving C_K as a function of W_D. Finally, the operating point P3 is reached representing the limiting case: hydraulic transport upward. The lost solids concentration C_K, respectively the clarification ratio β, follows by combining Eqs. (58) to (62):

$$C_K = \frac{W_0C_0 - W_D^{\frac{m-1}{m}}\left[(m-1)^{\frac{1-m}{m}} + (m-1)^{\frac{1}{m}}\right]}{W_0 - W_D}. \tag{66}$$

For other settling characteristics numerical methods are needed to solve the system of equations.

The thickening ratio n_E can be considerably increased, if the clarification ratio can be also increased. As demonstrated in Sect. 2.3.1.5 this can be effected by using lamella separators which have to be installed in the upper part of the flow tank.

The model has to be supplemented by further equations that describe the separation process in lamellae. It can be shown, that for only short parallel plates with slight slopes with respect to the vertical plane, significantly higher thickening ratios can be expected [34] if the settled particles that slide on the inclined bottoms of the plates are transferred into the compaction zone without resuspension.

In order to use these models it must be confirmed that the settling rate is only a function of solid concentration. Hence, different points discussed in Sect. 2.2.2.2 have to be considered in the evaluation of batch settling tests. In non-ideal suspensions like activated sludges, a significant compaction process can occur which is not taken into account by the model discussed.

Settling models that take both clarification and thickening into consideration can be combined with reactor models in order to describe the whole process — reaction, thickening of the produced biomass and reduction of the thickened biomass in the bioreactor [35,36,37].

2.3.2.6 *Experimental Results*

Unfortunately, until now only very few results for batch *and* continuous measurements have been published. One of this seldom reports is the paper of Dick and Young [33] that describes the sedimentation of activated sludge aerated with pure oxygen in a pilot plant. A vertical cylindrical flow tank with sludge scrapers and with a diameter of 2.5 m and approximately the same height was used. Figure 31 shows the measured settling characteristics.

In Fig. 34 theoretical solutions for four different dimensionless loads between $W_0C_0 = 4.3 \ldots 7.2$ and appropriate measuring points are represented. It can be

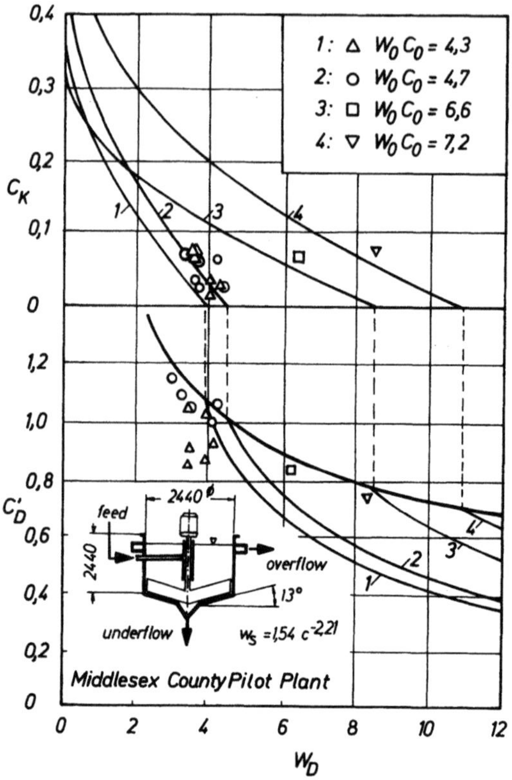

Fig. 34. Comparison of theoretical and experimental results for clarification and thickening of activated sludge in a vertical flow tank of a pilot plant (from Ref. [33])

seen that the operating staff tried to operate at the point "maximum load". The measured solid concentrations of the underflow are somewhat lower than the theoretical results. However, activated sludge is a non-ideal suspension. Therefore, the correlation of theoretical and experimental results is satisfactory.

2.4 Application of Sedimentation Processes in Biotechnology

Certainly, the main range of application is biological waste water treatment and sludge treatment because low operation costs are needed which characterize sedimentation processes. Table 4 shows some examples. In nearly all cases it is important to obtain both a very high clarification ratio in order to protect rivers and lakes and a high thickening ratio in order to reduce dewatering costs for the following solid-liquid separation processes (flotation, centrifugation, filtration). All types of sedimentation tanks are successfully used with the exception of lamella separators in which blockings and incrustations give reasons for shutdowns. Except for this field of application, sedimentation tanks are only used in continuous feed yeast production and in some new processes of continuous production of ethanol as an energy source (Table 5). However, until now this process has not reached commercial scale operation.

Table 4. Application of sedimentation tanks in biological waste water and sludge treatment

Process	Method, bioreactors	Separated microorganisms	Ref.
Waste water treatment by substrate removal nitrification or denitrification	Aeration of activated sludge in different reactor types	Activated sludge	[40] [41] [38]
	Trickling filters	Washed-out parts of biofilm	[38]
	Rotating biological contactors	Washed-out parts of biofilms	[38]
	Fluidized beds	Washed-out biological media from fluidized sand particles	[39]
Waste water treatment by anaerobic digestion	Digestion with sludge recycle (contact digestion)	Anaerobic bacteria	[42]
Sludge stabilization	Aerobic digestion with sludge recycle	Aerobic bacteria and death biological materials	[38]

Table 5. Application of sedimentation tanks in ethanol and feed yeast production

Product	Raw material, substrate	Separated microorganisms	Ref.
Ethanol	Hydrolyzed wood	Flocculent yeasts	[3]
Ethanol	Molasses from beets	Flocculent yeasts	[45]
Feed yeast	Waste from citric acid	Flocculent yeasts	[44]
Feed yeast	Waste from sauerkraut production	*Candida utilis*	[46]

3 Centrifugation

3.1 Centrifuges

3.1.1 General Remarks

The principle of operation consists in taking advantage of the relative motion of solid particles and liquid due to inertial forces in rotating suspensions. Two different basic types of centrifuges should be distinguished: filter centrifuges and sedimentation centrifuges.

In *filter centrifuges* an inner part of the rotating cylinder is formed by a filter media which is only permeable for the liquid flowing in radial direction. While the liquid is stored or drained in the gap between the filter media and the outer drum of the centrifuge, a filter cake is formed at the inner surface of the filter which has to be removed in a continuous or discontinuous way. Separation efficiency does not depend on density differences, but only on radial pressure gradients in filter cake and material. In Fig. 35 three different types of centrifuges are shown for continuous liquid flow. Only the pendulum centrifuge has to be stopped for removing the filter cake. In "pusher" and in "knife discharge" filter centrifuges the filter cake can be periodically removed during operation.

In *sedimentation centrifuges* the separation efficiency depends on density differences. Particles with higher densities move on spiral trajectories in radial and axial directions until they have reached the inner surface of the centrifuge drum. Therefore, the liquid has to be drained near the axis. In addition knife discharge- and double cone centrifuges with discontinuous or periodical solid discharge specially disc-, decanter- and tube centrifuges with continuous and discontinuous solid discharge are employed (Fig. 36). For reasons of their increasing importance for biochemical processes, they are explained in more detail.

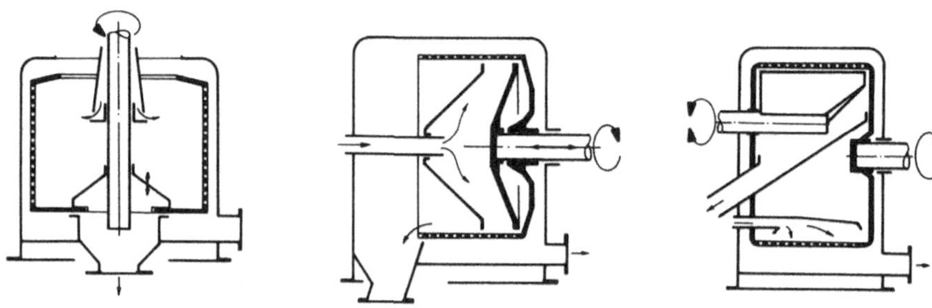

a) pendulum centrifuge b) pusher centrifuge c) knife centrifuge

Fig. 35a—c. Filter centrifuges

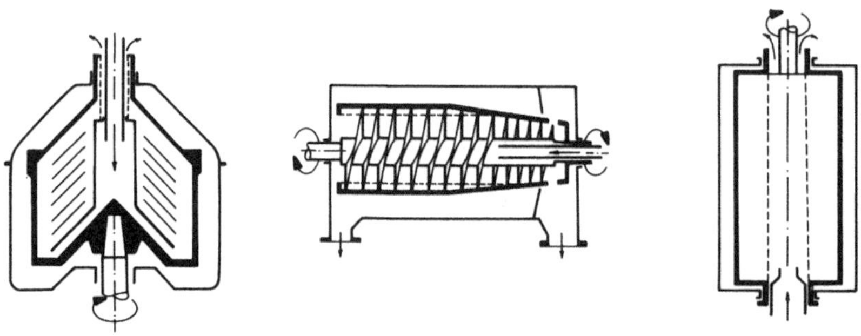

a) disc centrifuge b) decanter centrifuge c) tube centrifuge

Fig. 36a—c. Sedimentation centrifuges

3.1.2 Disc Centrifuges

As in the case of sedimentation due to gravity in lamella separators, settling distances can be reduced in centrifuges by using parallel plates of conical form.

Figure 36a shows a cross sectional drawing of a solids retaining disc centrifuge or plate separator. Within the drum, a number of conical plates rotate around a vertical axis. The suspension is fed from a central non-rotating feed tube and is guided outwards at the bottom of the drum. After flowing through cleavages into the conical spaces the suspension is separated. The liquid of lower density flows towards the axis, the settled solid particles move on the inner surfaces of the conical plates towards the drum. Nozzle discharge disc centrifuges are plate separators with continuous solid discharge. The solids are centrifuged by nozzles being normally installed at the highest diameter of the drum (Fig. 37a). For variable solid concentrations in the feed or for variable settling properties, it is more suitable to use separators with sludge return (Fig. 37b). In this way a poor thickening caused by lower solid concentrations in the feed can be diminished by an increase of sludge return. In separators for the treatment of biosuspensions with high viscosities the sludge can be discharged by lower pressure gradients and larger nozzles (Fig. 37c).

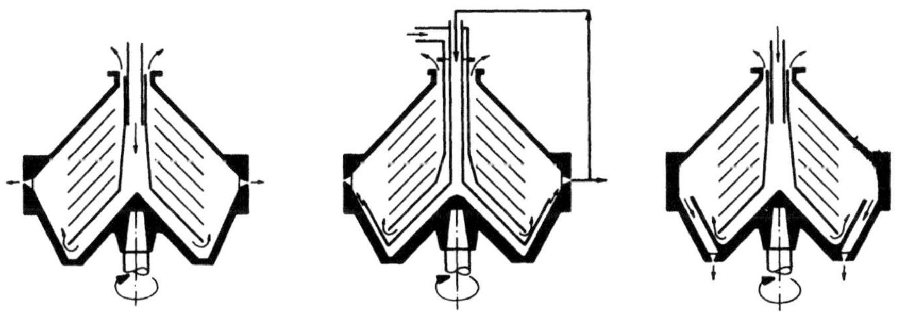

Fig. 37a—c. Disc centrifuges

3.1.3 Decanter Centrifuges

Decanter centrifuges have, contrary to separators, horizontal shafts with a solid jacket and a screw discharge for solids. They are especially suitable for continuous thickening of sludges with relative high solid concentration. The suspension is fed through an axially movable tube and flows through radial openings in the gap of the drum (Fig. 36b). The solid particles settle due to inertia forces on the inner surface of the rotating drum. A screw which rotates with a something lower number of revolution than the drum therefore transfers the solids from the cylindrical part into the conical part of the drum. Here the sludge is dewatered by compression which comes about by transport through decreasing cross sections (thickening zone). The fraction with the lower solid concentration flows through the cylindrical part of the drum in the other direction and leaves the centrifuge by flowing over a weir (clarification zone). Operation characteristics are mainly determined by the form of the drum. Decanters with a long cylindrical part and a short conical part of the drum are principally clarifiers. Decanters with the opposite relationship are mainly used as thickeners. Moreover, the separation process can be influenced by removal of the intake tube, by changing of the annular weir, by variation of the speed and the speed difference of drum and screw.

3.1.4 Tube Centrifuges

Tube centrifuges were developed for the use in laboratories or for the treatment of small industrial charges. Because of their relative small diameters they can work at high speeds, so that centrifugal accelerations of 50000 g can be reached. With such ultracentrifuges very small particles such as viruses and subcellular organelles can be separated. This kind of centrifuge is often equipped with cooling coils to remove the heat generated.

Figure 36c shows a simple tube centrifuge with a vertical axis. Frequently, rotors with 4 to 24 tube chambers for buckets of different sizes are used. In this cases, the tube axes are placed eccentricaly to the axis of rotation. Therefore, the solids are only settled at the external side of the tube chambers. In Fig. 38 two examples are presented with only a few large (Fig. 38a) or several small tube chambers (Fig. 38b) for the isolation of biological particles from a continuously flowing sample.

The following considerations are valid for separation in sedimentation centrifuges.

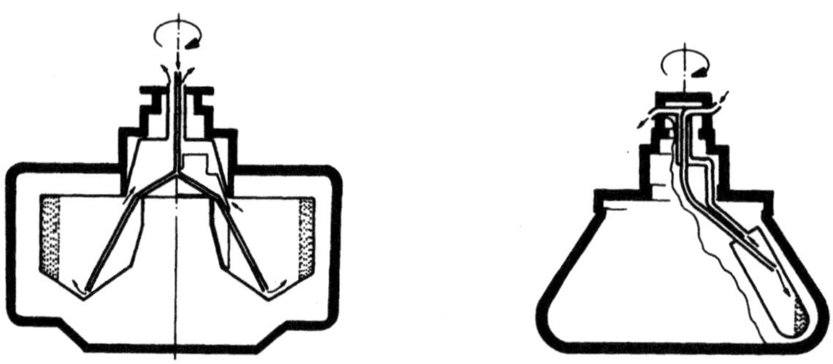

a) b)

Fig. 38a and b. Ultracentrifuges for different sizes and numbers of sample tubes (system SORVALL)

3.2 Centrifugation in Closed Systems

3.2.1 Centrifugation in the Range of Low Particle Concentrations

In a rotating suspension particles move on spiral trajectories. In closed systems the velocity of a particle has a tangential component w_θ and a radial component w_r. For the normal case of particle densities with a higher density than that of the liquid, the settling rate w_r is directed outward (Fig. 39). w_r is determinative for the rate of the separation process and takes the place of the settling rate due to gravity forces which can be neglected in centrifuges.

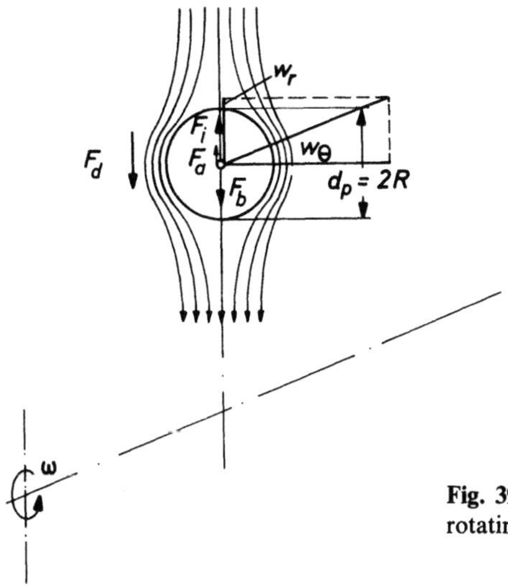

Fig. 39. Forces at a spherical particle, suspended in a rotating fluid (radial components)

The settling rate w_r of a spherical particle is influenced by the following forces (radial components):

the inertial or centrifugal force of the particle

$$F_i = \varrho_p V_p \frac{w_\Theta^2}{r} = \varrho_p V_p \omega^2 r \tag{67}$$

the inertial or centrifugal force of the displaced liquid (buoyant force)

$$F_b = \varrho V \omega^2 r \tag{68}$$

the drag force

$$F_d = \zeta \frac{\varrho w_r^2}{2A_p} \tag{5}$$

and the acceleration force

$$F_a = \varrho_p V_p \frac{dw_r}{dt} = \varrho_p V_p \frac{d^2r}{dt^2} \tag{69}$$

where

ω = angular velocity

r = distance between the mid-point of the particle and the axis of rotation

t = time

A_p = cross sectional area of the particle.

The sum of these forces becomes zero. The settling rate w_r increases with increasing r because centrifugal force increases. Therefore, the particle motion is always unsteady. It will be only allowed to neglect the acceleration force if the alteration of the coordinate r is small in comparison with the radius of the drum r_0. For this case w_r can be calculated in the same way as in Sect. 2.1.1.

Within the range of creeping flow (Re < 1), Stokes' law (Eq. (9)) can be again used to calculate the drag force. Then for w_r follows that

$$w_r = \frac{4}{18} R^2 \frac{g}{\nu} \left[\frac{\varrho_p}{\varrho} - 1\right] Z \tag{70}$$

with $Z = \frac{r_0 \omega^2}{g} = \frac{4\pi^2 r_0 n^2}{g}$ as the centrifugal acceleration number (71)

and $n = \frac{\omega}{2\pi}$ as the number of revolutions (72)

Substituting Eq. (11) in Eq. (70), we obtain

$$Z = \frac{w_r}{w_{s0}} \tag{73}$$

Centrifuge numbers of industrial centrifuges are of the order of 300 till 11000. Therefore, the volumes of centrifuges are much smaller than that of sedimentation tanks. With small tube centrifuges for use in laboratories, Z numbers of 50000 can be reached.

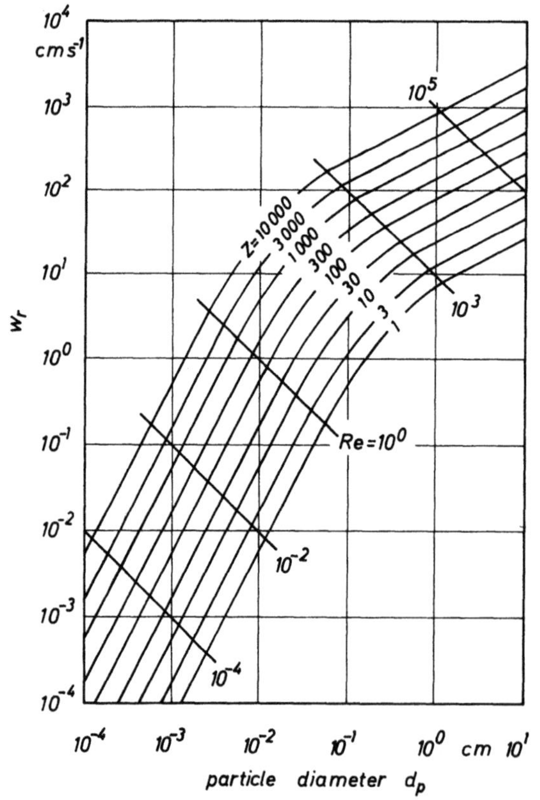

Fig. 40. Settling rates of single spherical particles in water for 10 °C and $\varrho_p/\varrho = 1{,}01$ (from Ref. [64])

For Re > 1 drag force has to be calculated from Eq. (12) or has to be taken from Fig. 7 by iteration methods. This range will be valid more frequently than in sedimentation due to gravity forces, because of the higher settling rate of the particles.

In Fig. 40 the settling rate w_r of spherical particles is presented as a function of particle diameter and the centrifugal acceleration number for a density ratio $\varrho_p/\varrho = 1.01$ and for water at 20 °C.

3.2.2 Centrifugation in the Range of High Particle Concentrations

Thickening caused by centrifugal forces can be studied by using transparent tubes in laboratory centrifuges and "stopping" their motion by stroboscopic lights, which is synchronized with the centrifuge (Fig. 41) [9]. In this way the slurry-liquid interface can be continuously observed as in batch sedimentation measurements (see Sect. 2.2.2.1).

Figure 42 shows typical sedimentation curves for a lime slurry measured in tubes with a volume of 10 ml. For high centrifugal acceleration numbers Z, the final solid

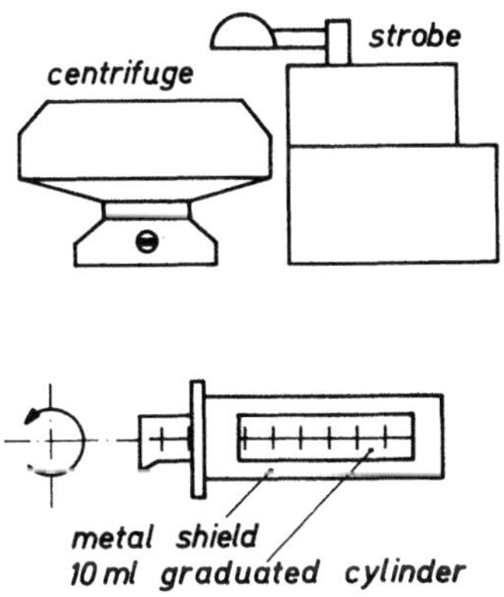

Fig. 41. Test tube centrifuge for the continuous observation of sludge settling [9]

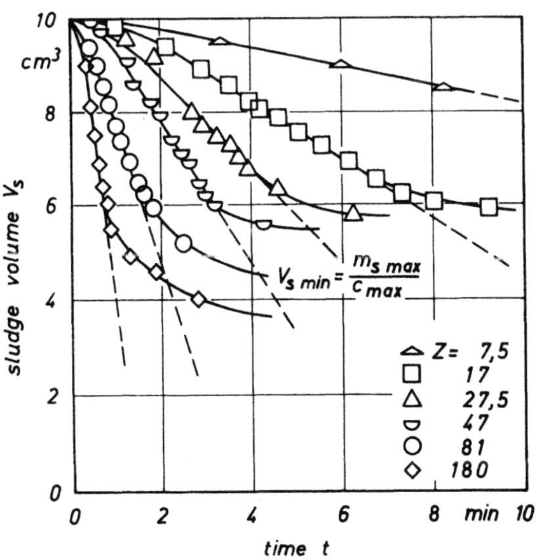

Fig. 42. Settling curves for a mixed digested sludge in a test tube centrifuge [9]; parameter: centrifugal acceleration number

concentration c_{max} can be reached in very short times. With increasing Z higher solid concentrations can be obtained. However, activated sludge does not compact to more than about 10% solids.

For a given centrifugal acceleration number, data from a series of these tests with different solid concentrations can be plotted as interface velocity versus solid concentration thus giving settling characteristics for centrifugation.

3.3 Centrifugation in Open Systems

3.3.1 Clarification in Tube Centrifuges

Until now clarification by centrifugal forces has only been approximately calculated for tube centrifuges with cylindrical drums and discontinuous solid discharge. The calculation method used is the same discussed in Section 2.3.1. It is based on the following assumptions:

1) Sedimentation rates of all particles are comparable and constant at all times
2) Fluid flow is of plug flow type
3) Particle concentration is constant in the inlet cross sectional area.

Figure 43 shows a section through a tube centrifuge and a "limiting trajectory", which starts at the point ($r = r_{0G}$, $z = 0$) and ends at the point ($r = r_s$, $z = L$). In reality the limiting trajectory is a three-dimensional spiral curve. The two-dimensional "limiting trajectory" is formed at all points of the spiral curve which are located in one section. Using Eq. (1) we obtain for the clarification ratio

$$\beta = \frac{\int_{r_{0G}}^{r_s} w_0 r \, dr}{\int_{r_i}^{r_s} w_0 r \, dr} \tag{74}$$

which is in fundamental conformity with Eq. (22) for sedimentation in horizontal flow tanks. It follows from Eq. (74) that

$$\beta = \frac{r_s^2 - r_{0G}^2}{r_s^2 - r_i^2} \tag{75}$$

because of the postulated plug flow condition.

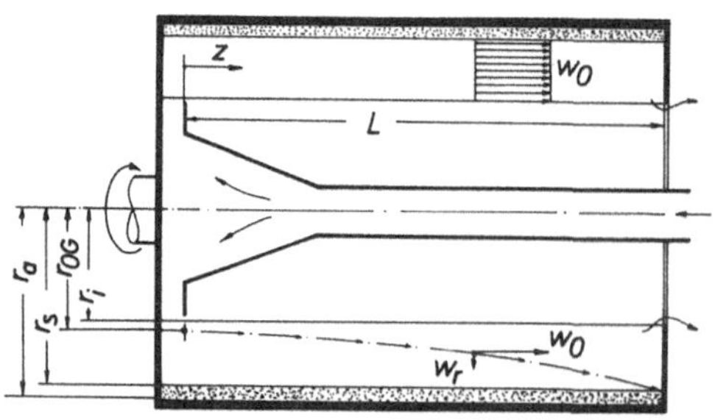

Fig. 43. Tube centrifuge with limiting trajectory

In Eq. (75) the initial height of the limiting trajectory is still unknown and can be estimated by solving the differential equation of the trajectories

$$w_0 dr = w_r dz \tag{76}$$

With

$$Z(r) = \frac{w_r}{w_s} = \frac{r\omega^2}{g} \tag{77}$$

Eq. (76) becomes

$$dz = \frac{w_0 g}{w_s \omega^2} \frac{dr}{r} \tag{78}$$

By integration of Eq. (78) and by considering the boundary conditions

$$\begin{aligned} r &= r_{0G} \quad \text{for} \quad z = 0 \\ r &= r_s \quad \text{for} \quad z = L \end{aligned} \tag{79}$$

r_{0G} can be calculated:

$$\ln \frac{r_s}{r_{0G}} = w_s \frac{\omega^2 L}{w_0 g} = \frac{w_s}{w_0} Z \frac{L}{r_s} \tag{80}$$

$$r_{0G} = r_s \exp\left[-\frac{L w_s Z}{w_0 r_s}\right] \tag{81}$$

Substituting Eq. (81) into (76) gives

$$\beta = \frac{1 - \exp\left[-\frac{2 L w_s Z}{w_0 r_s}\right]}{1 - r_i^2/r_s^2}. \tag{82}$$

For $\beta = 1$ and $r_s \approx r_a$ the condition for complete particle separation follows:

$$w_s = \frac{\dot{V}_0 r_a}{Z\, 2\pi L} \frac{\ln \frac{r_a^2}{r_i^2}}{r_a^2 - r_i^2} \tag{83}$$

where

$$\dot{V}_0 = w_0 \pi (r_a^2 - r_i^2) \tag{84}$$

In Eq. (83) the left hand side includes the separation properties of the suspension and the right hand side includes the properties of the centrifuge. Now, with a given suspension (w_{s1}) and a given tube centrifuge (r_{i1}, r_{a1}, L_1) the operational parameters Z_1 and $\dot{V}_{01}$ have to be estimated by experiments, so that particles will be completely separated. In order to estimate one of the following values of a second tube centrifuge ($\dot{V}_{02}$, Z_2, r_{i2}, r_{a2}, L_2) for the clarification of the same suspension ($w_{s1} = w_{s2}$), Eq. (85) can be used:

$$\frac{\dot{V}_{01} r_{a1}}{Z_1 L_1} \frac{\ln \frac{r_{a1}^2}{r_{i1}^2}}{r_{a1}^2 - r_{i1}^2} = \frac{\dot{V}_{02} r_{a2}}{Z_2 L_2} \frac{\ln \frac{r_{a2}^2}{r_{i2}^2}}{r_{a2}^2 - r_{i2}^2} \tag{85}$$

Sometimes this scale-up or scale-down method first proposed by Ambler [48] can also be applied to other types of centrifuges.

3.3.2 Clarification and Thickening in Decanter Centrifuges

The analogous system to sedimentation by gravity forces in vertical flow tanks would be a tube centrifuge with continuous radial solid discharge. However, the only centrifuges with both continuous liquid and solid discharges are the disc and the decanter centrifuges. For disc centrifuges no suitable calculation methods exist.

In decanter centrifuges liquids and solids are transferred in an axial direction. The solid flow rate is not fixed by the limiting flux as for vertical thickening, but by the transport capacity of the conveyer screw. For equal values of the mean residence time of thickened solids and the conveying rate, all solids are just discharged by the screw (Fig. 44) [9,49]

$$\frac{V_D}{\dot{V}_D} = \frac{L}{\Delta\omega\, s} \tag{86}$$

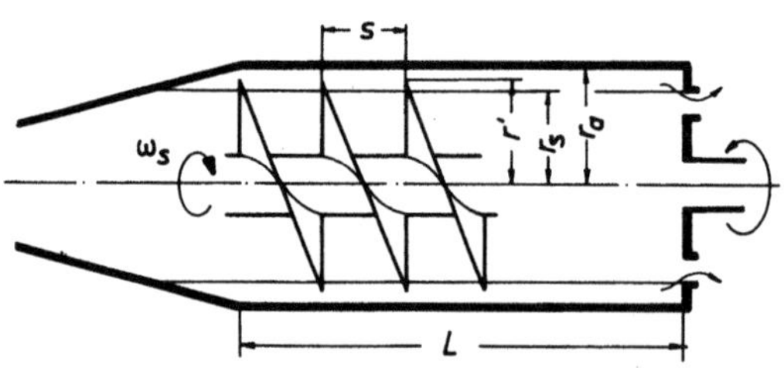

Fig. 44. Decanter centrifuge

where $\dot{V}_0$ = flow rate thickened solids
V_D = volume of thickened solids
$\Delta\omega = \omega - \omega_s$ = difference of the angular velocities of drum and screw
s = pitch of blades
L = length of the cylindrical part of the drum.

With the total mass balance

$$\dot{V}_0 c_0 = \dot{V}_D c_{max} \tag{87}$$

and

$$V_D \approx (r' - r_s)\, 2\pi r_a L \tag{88}$$

we obtain

$$\dot{V}_0 c_0 = (r' - r_s)\, 2\pi r_a \Delta\omega s c_{max} \tag{89}$$

Equation (88) is the condition for maximum load. For

$$\dot{V}_0 c_0 < (r' - r_s)\, 2\pi r_a \Delta\omega s c_{max} \tag{90}$$

the decanter is underloaded. More liquid is conveyed by the screw, and a smaller solid concentration $c_D < c_{max}$ is obtained because of the small solid feed or the high conveying rate. For

$$\dot{V}_0 c_0 > (r' - r_s)\, 2\pi r_a \Delta\omega s c_{max} \tag{91}$$

the decanter is overloaded. Indeed the maximum solid concentration c_{max} can be reached, but a part of the solids are discharged with the overflow so that clarification is affected ($\beta < 1$).

From Eq. (89) the needed difference in speed $\Delta n = \frac{\Delta\omega}{2\pi}$ for maximal load operation can be calculated. If c_{max} is unknown, with a given suspension (c_0, c_{max}) and a given decanter ($r_{s1}, r_1', r_{a1}, s_1$) the operational parameters $\dot{V}_{01}$ and $\Delta\omega_1$ have to be estimated by experiments for maximum load. To calculate the corresponding operational parameters $\dot{V}_{02}$ and $\Delta\omega_2$ of a second decanter ($r_{s2}, r_2', r_{a2}, s_2$) but with the same suspension, Eq. (92) can be used:

$$\frac{\dot{V}_{01}}{(r_1' - r_{s1})\, r_{a1}\, \Delta\omega_1\, s_1} = \frac{\dot{V}_{02}}{(r_2' - r_{s2})\, r_{a2}\, \Delta\omega_2\, s_2}\,. \tag{92}$$

This scale-up or scale-down method was first proposed by Veselind [9]. Equation (85) and (92) have to be applied if both clarification and thickening are to be considered.

3.4 Application of Centrifuges in Biotechnology

Centrifuges are of increasing importance for thickening of biological solid wastes that particularly arise from waste water treatment (Table 6). Obviously in this case decanter centrifuges have been proved to be effective to a higher degree than disc centrifuges. In other processes centrifuges have been used for many years to separate microorganisms from culture media. The most of these processes are discontinuous processes.

In Table 7 some continuous processes are specified. In these cases disc centrifuges have been more frequently used than decanter centrifuges.

Table 6. Application of centrifuges in biological waste water and sludge treatment

Process	Typ of centrifuge	Separated microorganism	Ref.
Waste water treatment by substrate removal denitrification	Disc	Activated sludge	50)
Thickening of excess sludge before anaerobic digestion	Decanter disc	Activated sludge	51) 52)
Thickening of mixed sludge after anaerobic digestion	Decanter	Digested organic solids and anaerobic bacteria	52) 53)
Thickening of mixed sludge before incineration	Decanter	Primary sludge and excess sludge (activated sludge)	54)
Thickening of aerobic mineralized sludge	Decanter	Aerobic bacteria and death biological materials	55)

Table 7. Application of centrifuges in different bioprocesses

Product	Raw material, substrate	Type of centrifuge	Separated microorganism	Ref.
Ethanol	Potatoes, grain	Disc or decanter	Yeasts	56)
SCP	n-Paraffin	Decanter	Yeast *Endomycopsis lipolytica*	57)
SCP	n-Paraffin	1. step: disc 2. step: decanter 1. step: flotation 2. step: decanter	Yeast Bacteria	58)
Feed yeast	Spent sulfite liquor	Disc	Yeast	43)
Baker's yeast	Melasse	Disc	Yeast *Saccharomyces cerevisiae*	43)
Amino acids	Krill	Disc	Chitinous exoskeleton and other solid materials	52)

4 Flocculation

Single microorganisms can be separated from culture media by filtration, flotation or centrifugation in centrifuges with relative high centrifugal acceleration numbers. Some recent results, published by Reuß et al. [16] show that sedimentation by gravity forces in lamella separators can be a further method to separate single yeast cells (Fig. 21). However, normally the formation of multi-cell groups (flocs) is necessary for biomass separation within an acceptable time or without the expenditure of too much energy.

Several microorganisms such as aerobic bacteria in the activated sludge or yeast cells used in brewing or SCP production flocculate, probably as a result of polysacharide molecules which are produced by the cell and are components of the capsular layer upon the outer cell wall (bioflocculation). A great number of fungi form pellets which show a higher density and a higher mechanical strength than flocs.

The formation of large dense flocs can be influenced by velocity gradients of the liquid. In the case of very small velocity gradients, the frequency of collisions of the particles is too low. Consequently, only small flocs can grow. With high velocity gradients the flocs are destroyed because the shear stress is higher than the mechanical strength. These facts have to be considered in the design and operation of flocculation reactors. A second step of flocculation can occur in sludge blankets of vertical flow tanks as a consequence of particle collisions. Therefore, very small particles are retained, which would otherwise escape.

Without floc formation a remarkably higher surface of sedimentation tanks or higher energy consumption of centrifuges would be needed.

In the case of insufficient bioflocculation, synthetic polyelectrolytes with molecular weights of 600 to 60000 can be added as organic flocculant aids. In general cationic types are used for destabilization of bacterial or algal suspensions. Both clarification and thickening can be improved by flocculation, however different types of polyelectrolytes will frequently have to be used if a high clarification ratio or a high thickening ratio are to be obtained. Polyelectrolytes must not injure the biological activity of cells. Harmful influences are only allowable with waste biomass. Detailed representations of flocculation principles and directions for technical applications are given by Atkinson and Daoud [60], Bratby [61], Aiba and Nagatani [62] and Wills [63].

5 Nomenclature

Symbols

A	surface of a sedimentation tank	L^2
a	retardation factor (see Eq. (21))	T
A_p	cross section area of a particle	L^2
B	width of a sedimentation tank	L
c	solid concentration	ML^{-3}
C	dimensionless solid concentration	—
CSV	Comparison Sludge Volume	—
d	diameter of settling tubes	L
d_p	diameter of a spherical particle	L
d'_p	coefficient in RRSB-distribution functions	L
F_g	force due to gravity	MLT^{-2}

F_b	buoyant force	MLT^{-2}
F_d	drag force	MLT^{-2}
F_a	acceleration force	MLT^{-2}
F_i	inertial force	MLT^{-2}
g	acceleration due to gravity	LT^{-2}
g'	coefficient in the frequency distribution function	—
H	height of liquid in a sedimentation tank	L
Hz	Hazen number (see Eq. (44))	—
h	height of the boundary between clear liquor zone A and zone B of collective subsidence	L
h_0	initial height of the suspension	L
k	coefficient in Eq. (56)	$LT^{-1}(ML^{-3})^n$
L	length of a sedimentation tank, of plates of a lamella separator or of a tube centrifuge	L
m	exponent in equations describing settling characteristics	—
n	number of revolutions	T^{-1}
n	exponent in RRSB-distribution functions	—
n_E	thickening ratio	—
Q	cumulative frequency distribution	—
q	frequency distribution	L^{-1}
R	radius of a spherical particle	L
Re	Reynolds number (see Eq. (8))	—
r	radius coordinate of a rotating particle	L
r_0	radius of a centrifuge drum	L
r_s	radius of a settled layer in a centrifuge drum	L
r_{0G}	radius of a starting point of a limiting trajectory	L
s	pitch of screw blades	L
SVI	Sludge Volume Index	L^3M
TS_R	dry solid matter	ML^{-3}
t	time	T
V	volume	L^3
V_p	particle volume	L^3
$\dot{V}$	flow rate	L^3T^{-1}
W	dimensionless fluid velocity	—
$\bar{W}_s$	dimensionless settling rate	—
w	fluid velocity	LT^{-1}
w_s	settling rate	LT^{-1}
w_{s0}	settling rate of single particles	LT^{-1}
x	cartesian coordinate	L
x_{0G}	vertical coordinate of a starting point of a limiting trajectory	L
y	cartesian coordinate	L
z	cartesian coordinate	L
Z	centrifugal acceleration number (see Eq. (73))	—
α	angle of inclination, slope of a lamella	—
β	clarification ratio	—

η	dynamic viscosity	$ML^{-1}T^{-1}$
ν	kinematic viscosity	L^2T^{-1}
ϱ	density of the fluid	ML^{-3}
ϱ_p	density of a particle	ML^{-3}
ζ	drag coefficient (see Eq. (5))	—
φ	solid flux	$ML^{-2}T^{-1}$
ω	angular velocity	T^{-1}

Indices

a	outer
i	inner
0	feed
B	zone B
C	zone C
D	zone D
K	overflow
D'	underflow
max	maximal
Σ	total
L	transport limiting layer

6 References

1. Stokes, G. G.: Trans. Camb. Phil. Soc. *9*, part II, 8 (1851)
2. Brauer, H.: Grundlagen der Einphasen- und Mehrphasenströmungen, p. 200, Aarau und Frankfurt a. M.: Verlag Sauerländer 1971
3. Comings, E. W., Pruiss, C. E., De Bord, C.: Ind. Eng. Chem. Process, Design and Develop. *46*, 1164 (1954)
4. Anderson, A. A., Sparkman, J. E.: Chem. Eng. *2*, 75 (1959)
5. Richardson, J. F., Zaki, W. N.: Chem. Eng. Sci. *3*, 65 (1954)
6. Coe, H. S., Clevenger, G. H.: Trans. Am. Inst. Min. Eng. *55*, 356 (1916)
7. Kynch, G. J.: Trans. Farad. Soc. *48*, 166 (1952)
8. Dick, R. I., Ewing, B. B.: J. of the Sanitary Eng. Division, Aug., 9 (1967)
9. Veselind, P. A.: Treatment and disposal of waste water sludges Ann Arbot Science Publ. Inc. Ann Arbor, Michigan, USA 1974
10. Eckenfelder, W. W., Melbinger, M.: Sewage and Industrial Wastes *29*, 1114 (1957)
11. Dick, R. I.: J. San. Eng. Div. ASCE *96* No. SA 2, Apr., 423 (1970)
12. Romagnoli, R. J.: Proc. of the 20th Ann. Industrial Waste Conf., p. 990 Purdue Univ., West Lafayette, Ind. May, 1974
13. Hibbert, R. L., Jones, W. F.: Water Pollution Control *73*, 14 (1974)
14. Götz, P.: Untersuchungen zur Eindickung von Belebtschlamm durch Sedimentation im Standzylinder, Studienarb. am Inst. für Chemieingenieurtechnik der TU Berlin, 1981
15. Johnstone, D. W. M., Rachwal, A. J., Hanbury, M. J.: Water Pollut. Control *78*, 337 (1979)
16. Reuß, M., Popovic, M., Jayanata, Y.: Fluiddynamische Probleme bei der alkoholischen Gärung, p. 65 4. Symp. Techn. Mikrobiologie, Berlin 1979
17. Hazen, A.: Americ. Soc. Civ. Eng., Paper 980, p. 45 (1904)
18. Wouda, T. W. M., Rietema, K., Ottengraf, S. P. P.: Chem. Eng. Sci. *32*, 351 (1977)

19. Binder, H.: Sedimentation aus Ein- und Mehrkornsuspensionen in schrägstehenden, laminar durchströmten Kreis- und Rechteckrohren, Dissertation TU-Berlin 1980
20. Camp, T. R.: Sewage Works Journal *8*, 742 (1936)
21. Jao, K. M.: JWPCF *42*, 220 (1970)
22. Pich, J.: Aerosol Science *3*, 351 (1972)
23. Binder, H., Wiesmann, U.: Chem.-Ing.-Techn. *52*, 332 (1980)
24. Oswald, P.: Untersuchungen zur Sedimentation von Hefen, Diplomarb., Inst. für Biotechn., TU-Berlin 1980
25. Walsh, T. J., Bungay, H. R.: Biotech. Bioeng. *21*, 1081 (1979)
26. Batel, W.: Einführung in die Korngrößenmeßtechnik, p. 16 Berlin: Springer 1964
27. Schmidt, M., Wiesmann, U.: Chem.-Ing.-Techn. *49*, 51 (1977)
28. Richtlinien für die Bemessung von einstufigen Belebungsanlagen mit Anschlußwerten über 10000 Einwohnergleichwerten, ATV-Regelwerte Abwasser, Arbeitsblatt A 131, Entwurf April 1980
29. Resch, A.: Untersuchungen an vertikal durchströmten Nachklärbecken von Belebungsanlagen, Berichte aus Wassergütewirtschaft und Gesundheitsingenieurwesen; Inst. für Bauingenieurwesen V Techn. Univ. München Nr. 29 (1981)
30. Dick, R. I.: J. Wat. Pol. Contr. Fed. *48*, 633 (1976)
31. Dick, R. I., Vesilind, P. A.: J. Wat. Pol. Contr. Fed. *41*, 1285 (1969)
32. Yoshioka, N.: J. Soc. Chem. Engng. *2*, 66 (1957)
33. Dick, R. I., Young, K. W.: Proc. 27. Ann. Ind. Waste Conf. Purdue Univ. Lafayette Ind. 1972
34. Binder, H., Putnaerglis, A., Wiesmann, U.: Numerical and experimental results for clarification and thickening sedimentation in vertical flow tanks, Vienna Euromech 144 (14.—16. 9. 81)
35. Mynhier, M. D., Grady, Jr., C. P. L.: J. Env. Engng. Div. ASCE *101*, 829 (1975)
36. Naito, M., Takamatsu, T., Fan, L. T,: Water Research *3*, 433 (1969)
37. Putnaerglis, A., Wiesmann, U.: Die Produktivität von Anlagen mit Biomassenrückführung bestehend aus Bioreaktor und Sedimentationsapparat, paper presented at the Sitzung des GVC-Fachausschusses Bioverfahrenstechnik Bad Dürkheim 25.–26. 5. 1981
38. Benefield, L. D., Randall, C. W.: Biological Process Design for Waste water Treatment, Prentice-Hall, Inc., Englewood Cliffs, N.J. 07632, 1980
39. Sutton, P. M. et al.: Oxitron System Fluidized Bed Waste Water Treatment Process: Development and Demonstration Studies; paper presented at the Joint Ann. Conf. of the Air Pollution Control Ass. on Pollution Control Ass. of Ontario, Toronto, Canada, April 1979
40. Pöpel, F.: Belebungsanlagen — Leistung, Berechnung, Entwurf — Deutscher Fachzeitschriften Verlag, Wiesbaden 1973
41. Barth, E. F.: Water Research *6*, 481 (1972)
42. Anderson, G. K., Donnelly, T.: New Processes of Waste Water Treatment and Recovery, p. 75 Chichester: Ellis Horwood, Ltd. 1978
43. Rehm, H.-J.: Industrielle Mikrobiol., p. 283 Berlin: Springer 1967
44. Braun, R. et al.: Process Biochem. *14*, 16 (1979)
45. Faust, U., Präve, P., Sukatsch, D. A.: Kontinuierliche Äthanolherstellung durch ein Gärverfahren der HOECHST/UHDE-Biotechnologie, p. 37 4. Symp. Techn. Mikrobiol., Berlin 1979
46. Hang, Y. D.: Process Biochem. *12*, 37 (1977)
47. Trawinski, H.: Zentrifugen und Hydrozyklone, Ullmanns Encyklopädie der Technischen Chemie, Vol. 2, p. 200, Weinheim/Bergstr.: Verlag Chemie 1972
48. Ambler, C. M.: Chem. Eng. Prog. *48*, 3 (1952)
49. Veselind, P. A.: J. Envir. Eng. Div. ASCE, 100 (1974)
50. BIOFUGAT®-Verfahren, information paper of the Wehrle Werk AG, Germany, Emmedingen
51. Tischer, W.: Abwassertechnik *19*, 34 (1978)
52. anonymus: Aufbereitungstechnik *18*, 493 (1977)
53. Birkholz, I., Lenz, G.: Korrespondenz Abwasser *25*, 158 (1978)
54. Becker, K. P., Wall, C. J.: Chem. Eng. Progress *72*, 61 (1976)
55. Zeper, J., Pepping, R.: Water Research *6*, 507 (1972)
56. Rosen, K.: Process Biochem. *13*, 26 (1978)
57. Birkenstaedt, J. W., Faust, U., Sambeth, W.: Process Biochem. *12*, 7 (1977)
58. Seipenbusch, R.: Verfahrenstechnische Probleme bei der Aufarbeitung von SCP, paper presented at the Sitzung des GVC-Fachausschusses Bioverfahrenstechnik, Stuttgart 1.—2. 4. 1976

59. Ellingsen, T., Mohr, V.: Process Biochem. *13*, 14 (1979)
60. Atkinson, B., Daoud, I. S.: Microbial Flocs and Flocculation in Fermentation Process Engineering, Adv. Biochem. Eng., Vol. 4, p. 41, Berlin: Springer 1976
61. Bratby, J.: Coagulation and Flocculation, Uplands Press Ltd., Croydon 1980
62. Aiba, S., Nagatani, M.: Separation of Cells from Culture Media, Adv. Biochem. Eng. Vol. 1, p. 31 Berlin: Springer 1971
63. Wills, R. F.: Sedimentation and Flocculation in Effluent Treatment, Biochemical and Biological Eng. Science, Vol. 1, p. 346 London: Academic Press 1967
64. Fair, G. M., Geyer, J. Ch., Okun, D. A.: Water and Wastewater Engineering Vol. 2: Water Purification and Wastewater Treatment and Disposal New York: John Wiley 1968
65. Trawinski, H.: Chem.-Ing.-Techn. *39*, 661 (1959)

Author Index Volumes 1–24

H. Engelhardt

High Performance Liquid Chromatography

Chemical Laboratory Practice

Translated from the German by G. Gutnikov
1979. 73 figures, 13 tables. III, 248 pages
ISBN 3-540-09005-3

Contents: Chromatographic Processes. – Fundamentals of Chromatography. – Equipment for HPLC. – Detectors. – Stationary Phases. – Adsorption Chromatography. – Partition Chromatography. – Ion-Exchange Chromatography. – Exclusion Chromatography. Gel Permeation Chromatography. – Selection of the Separation System. – Special Techniques. – Purification of Solvents. – Subject Index.

This simple and non-mathematical introduction to high-performance-liquid chromatography (HPLC) emphasizes the practical aspects of achieving a successful separation. This method usually permits analyses to be carried out more rapidly than by gas chromatography and is moreover, eminently suited for the separation of heatlabile, high-boiling, or nonvolatile substances, without lengthy or tedious derivatization. In principle, all substances that are stable in solution are amenable to separation by HPLC.
HPLC equipment is described in terms of the individual components, their expected performance capabilities and suitability for certain applications.
The areas of applications of the various separation techniques (adsorption, partition, ion-exchange, exclusion) are pointed out in order to facilitate selection of thc most appropriate technique by the worker for his particular problem. Considerable discussion is devoted to the parameters that are important in optimizing or improving a given separation.
The application of HPLC to actual problems in organic chemistry, pharmacological research, medicine, biochemistry and petrochemistry are illustrated by numerous relevant examples. This book is a translation of the well-known and very successful German edition.

Springer-Verlag
Berlin
Heidelberg
New York

GPSR Compliance

The European Union's (EU) General Product Safety Regulation (GPSR) is a set of rules that requires consumer products to be safe and our obligations to ensure this.

If you have any concerns about our products, you can contact us on ProductSafety@springernature.com

In case Publisher is established outside the EU, the EU authorized representative is:

Springer Nature Customer Service Center GmbH
Europaplatz 3
69115 Heidelberg, Germany

Batch number: 10620881

Printed by Printforce, the Netherlands